Sustainable Development Goals Series

The **Sustainable Development Goals Series** is Springer Nature's inaugural cross-imprint book series that addresses and supports the United Nations' seventeen Sustainable Development Goals. The series fosters comprehensive research focused on these global targets and endeavours to address some of society's greatest grand challenges. The SDGs are inherently multidisciplinary, and they bring people working across different fields together and working towards a common goal. In this spirit, the Sustainable Development Goals series is the first at Springer Nature to publish books under both the Springer and Palgrave Macmillan imprints, bringing the strengths of our imprints together.

The Sustainable Development Goals Series is organized into eighteen subseries: one subseries based around each of the seventeen respective Sustainable Development Goals, and an eighteenth subseries, "Connecting the Goals", which serves as a home for volumes addressing multiple goals or studying the SDGs as a whole. Each subseries is guided by an expert Subseries Advisor with years or decades of experience studying and addressing core components of their respective Goal.

The SDG Series has a remit as broad as the SDGs themselves, and contributions are welcome from scientists, academics, policymakers, and researchers working in fields related to any of the seventeen goals. If you are interested in contributing a monograph or curated volume to the series, please contact the Publishers: Zachary Romano [Springer; zachary.romano@springer.com] and Rachael Ballard [Palgrave Macmillan; rachael.ballard@palgrave.com].

Fraya Frehse · Angela Million ·
Ignacio Castillo Ulloa
Editors

Spatial Methods in Transdisciplinarity for Urban Sustainability

A Transformative Methodological Spectrum

Editors
Fraya Frehse
Department of Sociology + SMUS
University of São Paulo
São Paulo, Brazil

Angela Million
Department of Urban and Regional Planning + SMUS
Technische Universität Berlin
Berlin, Germany

Ignacio Castillo Ulloa
Department of Urban and Regional Planning + SMUS
Technische Universität Berlin
Berlin, Berlin, Germany

ISSN 2523-3084   ISSN 2523-3092 (electronic)
Sustainable Development Goals Series
ISBN 978-3-031-84366-2   ISBN 978-3-031-84367-9 (eBook)
https://doi.org/10.1007/978-3-031-84367-9

Color wheel and icons: From https://www.un.org/sustainabledevelopment/
Copyright © 2020 United Nations. Used with the permission of the United Nations.

The content of this publication has not been approved by the United Nations and does not reflect the views of the United Nations or its officials or Member States.

© The Editor(s) (if applicable) and The Author(s) 2025
This book is an open access publication.

Open Access This book is licensed under the terms of the Creative Commons Attribution 4.0 International License (http://creativecommons.org/licenses/by/4.0/), which permits use, sharing, adaptation, distribution and reproduction in any medium or format, as long as you give appropriate credit to the original author(s) and the source, provide a link to the Creative Commons license and indicate if changes were made.
The images or other third party material in this book are included in the book's Creative Commons license, unless indicated otherwise in a credit line to the material. If material is not included in the book's Creative Commons license and your intended use is not permitted by statutory regulation or exceeds the permitted use, you will need to obtain permission directly from the copyright holder.
The use of general descriptive names, registered names, trademarks, service marks, etc. in this publication does not imply, even in the absence of a specific statement, that such names are exempt from the relevant protective laws and regulations and therefore free for general use.
The publisher, the authors and the editors are safe to assume that the advice and information in this book are believed to be true and accurate at the date of publication. Neither the publisher nor the authors or the editors give a warranty, expressed or implied, with respect to the material contained herein or for any errors or omissions that may have been made. The publisher remains neutral with regard to jurisdictional claims in published maps and institutional affiliations.

This Springer imprint is published by the registered company Springer Nature Switzerland AG
The registered company address is: Gewerbestrasse 11, 6330 Cham, Switzerland

If disposing of this product, please recycle the paper.

Funding Information

This work was supported by the Global Center of Spatial Methods for Urban Sustainability (SMUS), which is funded by the German Federal Ministry for Economic Cooperation and Development (BMZ) via the German Academic Exchange Service (DAAD)—Project Nr. 57526630.

Summary

The book critically addresses the role of spatial methods in a transdisciplinary research-practice agenda regarding the promotion of urban sustainability throughout the globe with the aid of eight different transdisciplinary approaches primarily based throughout the Global South and jointly penned by academics and practitioners. While the range of methodological discussions regarding research-and-practice collaborations between scientific researchers and local practitioners (based in NGOs, private firms or local government agencies) as well as independent policy-makers or artists for the purpose of urban sustainability has been thriving vastly over the last years, little attention has been paid to *spatial* methods in particular. This is not to mention their *transdisciplinary* use in urban contexts of the *Global South*. Resorting on empirical settings as diverse as Durban (South Africa), Porto Alegre and São Paulo (Brazil), Kolkata (India), Bangkok (Thailand) and Tshwane (South Africa) during the last four (partially Covid-19 pandemic) years as well as Tangerang (Indonesia) from 2001 to 2021, and San José (Costa Rica) between 2004 and 2007, the book sheds light on the following, twofold question: *Which possibilities and limitations can spatial methods respectively unravel and encounter for transdisciplinary research and practice, in view of the SDG11 targets?*

The Editors

About the Editors and Contributors

Editors

Fraya Frehse Department of Sociology + SMUS, University of São Paulo, São Paulo, Brazil

Angela Million Department of Urban and Regional Planning + SMUS, Technische Universität Berlin, Berlin, Germany

Ignacio Castillo Ulloa Department of Urban and Regional Planning + SMUS, Technische Universität Berlin, Berlin, Germany

Editorial Assistant

Paulo Scarpa Ó Editorial, São Paulo, Brazil

Contributors

Shreyashi Bhattacharya Rekhi Centre of Excellence for the Science of Happiness – Indian Institute of Technology Kharagpur, Kharagpur, India

Ignacio Castillo Ulloa Department of Urban and Regional Planning + SMUS, Technische Universität Berlin, Berlin, Germany

Jacqueline Custódio Department of Urban Planning, Federal University of Rio Grande do Sul, Porto Alegre, Brazil

Jacques du Toit Department of Town and Regional Planning, University of Pretoria, Pretoria, South Africa

Fraya Frehse Department of Sociology + SMUS, University of São Paulo, São Paulo, Brazil

Sasidulal Ghosh State Fisheries Development Corporation Limited – West Bengal & Leaseholder of Jhagrashisha Bheri – East Kolkata Wetlands, Kolkata, India

Ilan Guest SATPLAN ALPHA, Johannesburg, South Africa

Raldi Hendro Koestoer School of Environmental Science, Universitas Indonesia, Jakarta, Indonesia

Budi Heru Santosa National Research and Innovation Agency (BRIN), Jakarta, Indonesia

Karina Landman Department of Town and Regional Planning, University of Pretoria, Pretoria, South Africa

Inês Martina Lersch Department of Urban Planning + SMUS, Federal University of Rio Grande do Sul, Porto Alegre, Brazil

Sandile Mbatha Department of Cooperative Governance and Traditional Affairs [COGTA], Pretoria, South Africa

Angela Million Department of Urban and Regional Planning + SMUS, Technische Universität Berlin, Berlin, Germany

Jenia Mukherjee Department of Humanities and Social Sciences + SMUS, Indian Institute of Technology Kharagpur, Kharagpur, India

Giulia Pereira Patitucci Federal Ministry of Management and Innovation in Public Services, Brasília, Brazil

Amy Pieterse Smart Places Cluster – South African Council for Scientific and Industrial Research, Pretoria, South Africa

Caio Moraes Reis Department of Sociology + SMUS, University of São Paulo, São Paulo, Brazil

Jakkrit Sangkhamanee Department of Sociology and Anthropology + SMUS, Chulalongkorn University, Bangkok, Thailand

Piyathep Tanmahasmut Chulalongkorn University, Bangkok, Thailand

Poramin Watnakornbancha Ari Ecowalk, Bangkok, Thailand

About the Contributors

Shreyashi Bhattacharya is a Senior Research Fellow of the Rekhi Centre of Excellence for the Science of Happiness at the Indian Institute of Technology Kharagpur, where she concluded her PhD on socio-ecologies of the Adi Ganga riverine system by addressing the history and resilience of the Adi Ganga, a heritage river flowing through the dense urbanscapes of Kolkata. In this study she used the transdisciplinary lens of 'ethno-graphy' to study the river's 'hydrosociality'—historically complex and dynamic socio-ecological assemblages. She has received the prestigious Writing Urban India Fellowship, Huntington Fellowship and Living Waters Museum Fellowship for her doctoral thesis.

Ignacio Castillo Ulloa, PhD is a Lecturer and Researcher at the Chair of Urban Development and Urban Design, Department of Urban and Regional Planning, and scientific coordinator of the Global Center of Spatial Methods for Urban Sustainability (SMUS), Technische Universität Berlin. He is co-editor of the book *Spatial Transformations* published by Routledge. He is also a research associate at the Collaborative Research Centre 1265 'Re-Figuration of Spaces'. His research interests include socio-spatial development and alternative disruptive (local) practices, critical urban research, transdisciplinarity and the use of spatial research methods.

Jacqueline Custódio is a lawyer who specialises in Public Law and a member of ICOMOS Brazil. She holds an MA in Museology and Heritage from the Federal University of Rio Grande do Sul, where she is currently pursuing her PhD in Urban and Regional Planning (PROPUR/UFRGS) with a focus on participatory planning and cultural heritage in the city of Porto Alegre. She is a member of the Sectorial Collegiate of Material Heritage at the National Council of Cultural Policy of Brazil's Ministry of Culture (CNPC/MinC). She worked as a member of both the Rio Grande do Sul State Council of Culture (CEC/RS) and the Sectorial Collegiate of Memory & Heritage of the Porto Alegre Municipal Council of Culture.

Jacques du Toit holds a PhD in methodology and is an Associate Professor in the Department of Town and Regional Planning, University of Pretoria, South Africa. His substantive research interest is in urban sustainability, particularly water-sensitive planning and design, and his procedural research interest is in evidence-based urban planning. He teaches infrastructure planning, policy analysis and research methods, and coordinates final year research report and mini-dissertation projects. Part time he also conducts applied research and serves on various institutional and faculty level committees.

Fraya Frehse is Professor of Sociology at the University of São Paulo, where she coordinates the Center for Studies and Research on the Sociology of Space and Time (NEPSESTE) and acts as a Lead Partner and Action Speaker of SMUS. She is an alumna of the Alexander von Humboldt Foundation and life member of Clare Hall College (University of Cambridge). Her research focuses mainly on urban theory; body, public space and urbanisation; social inequality/poverty and urban (public) space; homelessness; intersectionality and space; space and time in sociology; urban sustainability and public space; cultural heritage; urban visual culture; sociology of everyday knowledge.

Sasidulal Ghosh is the leaseholder of Jhagrashisha Bheri in the East Kolkata Wetlands. He was the former Deputy Director of the Department of Fisheries at the Government of India. He is passionate about inland fishing experimentation and sewage-fed fisheries. He is also a representative of the East Kolkata Wetlands Management Authority and has been a key member in the Wastewater User Association and Fish Producers' Association.

Ilan Guest is a professional town planner and Geographic Information Systems (GIS) specialist focused on spatial policy, rural development and online GIS applications. He concluded his MA in City and Regional Planning at the Georgia Institute of Technology in Atlanta Georgia (USA) as a Fulbright Scholar with specialisation in GIS and Remote Sensing. He is also an external lecturer at the Town Planning Department at the University Pretoria, where he teaches topics around regional and rural development, and the role of GIS and spatial data applications for the planning profession.

Raldi Hendro Koestoer is a faculty member at the School of Environmental Science, Universitas Indonesia. He holds BA & Doktorandus in Geography from Universitas Indonesia, an MA in Regional Science from the University of Queensland and a PhD in Environmental Science from Griffith University, Australia. Having been a Professor of LIPI, he has held various government positions, including Indonesian Senior Official for IMT GT and BIMP EAGA (2009–2018) and Principal Senior Policy Analyst (2018 to present). His research interests encompass urban and regional environments, spatial planning, sustainable development, spatial modelling and public policy.

Karina Landman is a Professor and HOD in the Department of Town and Regional Planning at the University of Pretoria with a background in Architecture and Urban Design. Her work focuses on sustainable development, including resilience and regenerative sustainability and public space.

Inês Martina Lersch is Professor of Urban Planning at the Faculty of Architecture of the Federal University of Rio Grande do Sul. She has a BA in Architecture and Urban Planning

(1998) from the same University, where she also completed her MA in Civil Engineering (2003) and her PhD in Urban and Regional Planning (2014). Her work primarily focuses on urban planning, planning history and heritage at the Postgraduate Program in Urban and Regional Planning (PROPUR/UFRGS). She is a SMUS Post-Doctoral Researcher (2022) and also a SMUS Partner.

Sandile Mbatha is the National Chief Data Officer at the Department of Cooperative Governance and Traditional Affairs, South Africa, where he is responsible for establishing the National Strategic Hub. The core of his work is to integrate, transform, analyse and visualise country-level data to extract value for informing strategic decision-making on planning, resource allocation, implementation and monitoring. He is responsible for building the National Strategic Hub which is the country's single source of truth. Previously he served as Senior Manager, responsible for Data, Research and Policy Advocacy in eThekwini Municipality where he supported evidence-based policy development processes through city-wide research and strategic planning.

Angela Million née Uttke, is Professor for Urban Design and Urban Development at TU Berlin, Germany. She is Director of SMUS. Her most current research explores educational landscapes, neurourbanism, multifunctional infrastructure as well as hybrid spatial constructions of young people within the Collaborative Research Center 1265 'Refiguration of Spaces'. She is founding member of JAS (Jugend Architektur Stadt e.V.), a non-profit association dedicated to built-environment education for young people. As an urban planner and urban designer, she is part of design juries and councils in different German cities.

Jenia Mukherjee is an Associate Professor of History of Ecology and Environment at the Department of Humanities and Social Sciences of the Indian Institute of Technology Kharagpur. Her interest lies in urban studies and transdisciplinary waters. In the book *Blue Infrastructures*, she proposed and applied historical urban political ecology (HUPE) to explore the more-than-contested wastewaterscapes of Kolkata, discussing its viability in forging academia-practitioners dialogues and actions. Recipient of the World Social Science Fellowship (2013), the Carson Writing Fellowship (2018) and the Nippon Foundation Fellowship (2020) for advancing urban ecological research, she is an active SMUS partner, investigating multiple projects from India. Together with Shreyashi Bhattacharya, Mukherjee has experimented with the efficacy of the 'ethno-graphic' approach in capturing (urban) environments of South Asia. Their latest output on 'ethnography' is available as a Springer video from 2023.

Giulia Pereira Patitucci is an architect and urban designer trained at the Faculty of Architecture and Urbanism of the University of São Paulo (2017), where she completed her MA on the issue of housing in the lives of homeless people (2022). Between 2018 and 2022, she worked with public policies for homeless population at the São Paulo Municipal Secretariat for Human Rights and Citizenship. In 2023, she continued addressing the same issue as coordinator of the Executive Secretariat for Strategic Projects at the São Paulo City Hall before moving to Brasília to become project manager of the Programme for the Democratisation of Union Properties at the Ministry of Management and Innovation in Public Services from the Federal Government of Brazil in early 2024. In 2017, she worked at the Social Rent Program of the Sao Paulo Metropolitan Housing Company.

Amy Pieterse holds an MA in Town and Regional Planning from the University of Pretoria and is a Senior Knowledge Applicator in the Smart Places Cluster at the Council for Scientific and Industrial Research, South Africa. Her expertise includes climate change adaptation planning, mainstreaming, policy analysis and local government planning. She has a specific research interest in how local governments use research and evidence in planning and policy, with a focus on adaptation. She is an innovator, developer and manager of the GreenBook (greenbook.co.za), a practical online planning support for the development of climate-resilient South African cities and towns.

Caio Moraes Reis is a SMUS-funded PhD candidate in Sociology at the Faculty of Philosophy, Languages and Literature, and Human Sciences of the University of São Paulo (FFLCH-USP), where he completed his MA in Political Science (2019) and his BA in

Social Sciences (Sociology, Anthropology, and Political Science) (2016). He is a researcher at the USP Center for Studies and Research in Sociology of Space and Time (NEPSESTE) and at SMUS. His research focuses on homelessness, poverty, death and urban sustainability.

Jakkrit Sangkhamanee is an Associate Professor of Anthropology at the Chulalongkorn University Faculty of Political Science in Bangkok. His work focuses on Science and Technology Studies, specifically hydrological engineering projects related to the Thai state formation and climate politics. His latest publications include 'An Assemblage of Thai Water Engineering: The Royal Irrigation Department's Museum for Heavy Engineering as a Parliament of Things' (*Engaging Science, Technology and Society*, vol. 3, 2017), 'Infrastructure in the Making: The Chao Phraya Dam and the Dance of Agency' (*TRaNS: Trans-Regional and -National Studies of Southeast Asia*, 6(1), 2018) and 'Bangkok Precipitated: Cloudbursts, Sentient Urbanity, and Emergent Atmospheres' (*East Asian Science, Technology and Society: An International Journal* (*EASTS*), 15(2), 2021).

Budi Heru Santosa specialises in water resource management, with a focus on flood risk management. He employs an interdisciplinary and transdisciplinary approach, integrating scientific analyses from various fields and non-academic knowledge. Additionally, he leverages advanced technologies such as Artificial Intelligence, Geographic Information Systems (GIS) and remote sensing to assess changes in river basin flood susceptibility resulting from land use and spatial planning policies. Currently, he serves as Chair of the Indonesian National Committee for UNESCO Intergovernmental Hydrological Programme.

Piyathep Tanmahasmut is an architectural and design anthropologist who worked with urban planners and architects in Bangkok's CBD and cultural areas before starting to cooperate with the Bangkok Teak Research by using ethnographic methodology to develop urban solutions for affordable housing, multi-generational homes and mixed-use spaces. He is also an MA student at Chulalongkorn University, where he pursues his interests in urban settlements, Anthropocene, and urban assemblage. His research focuses on shophouse residents in Chinese Thai neighbourhoods facing gentrification.

Poramin Watnakornbancha is a marketer and a naturalist. He has been working on the Ecowalk initiative, a place-based learning program aiming to advocate human-nature connection in the realm of biodiversity, since 2019. His works focus on collecting Ari-Pradiphat's biodiversity data, addressing environmental injustice and promoting benefits of urban nature ecosystem services. He is keen on utilising data derived from Ecowalk devising Bangkok's green policies. Amidst Climate crisis, the cities are his chosen battlefield for change.

Contents

Introduction: Advancing Transdisciplinarity for Urban
Sustainability Through Spatial Methods xxi

INTRODUCCIÓN: Transdisciplinariedad avanzada para la
sostenibilidad urbana a través de métodos espaciales xxxi

Part I   The Critical Role of Transdisciplinarity Through Spatial Methods

1 **A Methodological Framework for Transdisciplinary Urban Planning** 3
  Jacques du Toit, Amy Pieterse, and Sandile Mbatha
  1.1 Introduction 3
  1.2 A Review of Literature on Transdisciplinarity
       for Urban Sustainability 4
       1.2.1 Examples of Transdisciplinarity for
              Urban Sustainability......................... 4
       1.2.2 Principles for a Methodological Framework 7
  1.3 A Methodological Framework for Transdisciplinary Urban
       Planning 7
  1.4 Practitioner Reflections on the Framework Using the
       Example of Planning Support Science (PSSci) and a
       Customised Planning Support System (PSS) 11
       1.4.1 The Concept of PSSci and PSS 11
       1.4.2 eThekwini GreenBook MetroView 11
  1.5 Conclusion 14
  References.. 15

2 **Incremental, Iterative, Transformative: A
  Social-Learning Approach in Spatial and Transdisciplinary
  Research and Practice** 19
  Ignacio Castillo Ulloa
  2.1 Seeing Spatial and Transdisciplinary Research Through
       'Interrelated Strings' of Social Learning 19

	2.2	Delineating the Theoretical Purview	19
		2.2.1 Circumventing Ambiguous Rhetoric: Definitional Boundaries of Social Learning	19
	2.3	Sharpening the Focal Take: Delineating the Spatiality and Transdisciplinarity of Research and Practice	20
		2.3.1 Untying Multi-, Inter- and Transdisciplinarity Knots	20
		2.3.2 Transdisciplinarity in Practice—An Importance Nuance	22
		2.3.3 Methods Are Not Spatial but Rather Enable to Fathom Out Phenomena Spatially	23
	2.4	Telescoping the Gist of Social Learning's Instrumentalising within Specific Instances of Research and Practice	24
		2.4.1 Social Learning as a Planning Practice and Urban Development Catalyser	24
		2.4.2 Rendering Transdisciplinarity Transformative	25
		2.4.3 Social Learning as the Building Block of Sustainability Transitions	26
		2.4.4 Overcoming Technocracy and Furthering Sustainability Through Action-Oriented Knowledge	27
		2.4.5 Social Learning Is Not the Bullseye—But Rather What it Enables	27
	2.5	Will-o'-the-wisp: The Pieces May Be at Hand but Won't Necessarily Fit	28
	2.6	Partition and Addition: The Underlying Logic of Master Planning Through *Planes Reguladores*	29
	2.7	Muddling Through Fixed Roles and Duties: Attempting to Surpass Mapping's Inherent Reductionism	30
	2.8	Missing the Forest for the Trees: The Consequences of Disciplinary Overspecialisation and Task Compartmentalisation	32
	2.9	From Talking the Talk to Walking the Walk: Some Reflections	33
	References		34
3	**Participation in Transdisciplinary Urban Planning Practice and Research: Spatial Methods in Action**		**37**
	Inês Martina Lersch, Angela Million, and Jacqueline Custódio		
	3.1	Introduction: Linking the World of Participation and Transdisciplinarity for SDG11	37
	3.2	Participation in Transdisciplinary Research and Practice	38
	3.3	Participation in Urban Planning Pedagogy and the Role of Spatial Methods	40
	3.4	Spatial Methods in Participatory Transdisciplinary Research: Historic Waterfront Development in Porto Alegre, Brazil	42

		3.4.1	Porto Alegre's Cais do Porto in the Context of SDG11	43
		3.4.2	Analysis of Spatial Methods in a Dynamic Participatory and Transdisciplinary Process	46
		3.4.3	Discussion of the Case Study, Porto Alegre, Brazil	47
	3.5	Conclusion: Transdisciplinary and Participatory Integration Between Planning Pedagogy, Research and Practice		52
	References...			53

Part II The Critical Role of Spatial Methods in Transdisciplinarity

4 Urban Sustainable Interactions by Homeless People Here and Now via Spatial Methods 59
Fraya Frehse, Caio Moraes Reis, and Giulia Pereira Patitucci
- 4.1 An Introduction to Homelessness and SDG11............. 59
- 4.2 A Counterintuitive Issue................................ 60
- 4.3 Contextualising the SMUS Toolkit...................... 61
- 4.4 Approaching Urban-Sustainable Homelessness via the SMUS 'Glasses' 65
  - 4.4.1 Present-Day, Environmentally Sustainable and Inclusive Public Spaces 66
  - 4.4.2 One Future Prospect of Socially Inclusive Public Spaces.................................. 69
- 4.5 Interactional Lessons Learned 71
- References... 74

5 Ethno-graphy on the East Kolkata Wetlands: A Transformative, Transdisciplinary Tool in Protecting Urban Ecological Heritage............................... 77
Jenia Mukherjee, Shreyashi Bhattacharya, and Sasidulal Ghosh
- 5.1 Introduction 77
- 5.2 The East Kolkata Wetlands............................ 79
- 5.3 Project Design and Team Composition 81
- 5.4 Phase-Wise Project Execution 82
  - 5.4.1 Transecting (Un)familiar Trails: 'Ethno-graphic' Training, Exchange and Exposure 82
  - 5.4.2 Selections and Combinations of Methods with Narratives Taken as Agents of Change 86
  - 5.4.3 The Trained as Trainers: Academia-School-Practitioner Engagement Through 'Ethno-graphy' 88
- 5.5 PEIP-ing Through the SMUS Lens: (Re)adjustments and Reinvigorations 90
- 5.6 Ethno-graphy as a Participatory, Agency-Induced Disseminative Tool.................................. 92
- 5.7 Conclusion 94
- References... 96

6	**Hybrid Use of Spatial Methods in Transdisciplinary Urban Sustainability Studies: Perspectives from Bangkok**...........	99
	Jakkrit Sangkhamanee, Piyathep Tanmahasmut, and Poramin Watnakornbancha	
	6.1 Bangkok Urban Resilience and Spatial Methods...........	101
	6.1.1 Geographic Information Systems (GIS)	101
	6.1.2 Spatial Analysis.............................	103
	6.1.3 Ethnography................................	105
	6.2 The Usefulness of Spatial Methods for Urban Sustainability Studies	107
	6.3 Sensorial and Affective Spatial Methodology: An Ongoing Experiment from Ari Ecowalk	108
	6.4 Conclusion	111
	References..	112
7	**Implementing a Transdisciplinary Approach in Flood Risk Management: Insights from Tangerang, Indonesia**...........	117
	Raldi Hendro Koestoer and Budi Heru Santosa	
	7.1 A Path Towards Resilient Flood Management: Transdisciplinarity, Systemic, Community-Based and Spatial	117
	7.2 Making Perceptions Spatially Tangible: Bringing the Spatial Dimension of Risk to the Fore	119
	7.2.1 Study Areas	119
	7.2.2 River Basin Flood Susceptibility Analysis..........	119
	7.2.3 Community Flood Resilience Analysis: Listening to Those Who 'Are in the Same Boat'	121
	7.3 Results and Discussion...............................	122
	7.3.1 Participatory Spatial River Basin Flood Susceptibility	122
	7.3.2 Participatory Community Flood Resilience Analysis	123
	7.3.3 Transdisciplinary in Flood Resilience Study........	125
	7.4 Conclusion	127
	References..	128
8	**Bridging the Gap Between Academia, Practitioners and Communities: A Transdisciplinary Process Towards Regenerative Public Space in South Africa**	131
	Karina Landman and Ilan Guest	
	8.1 Introduction	131
	8.2 The Evolution of Sustainability Thinking and Its Implications for Transdisciplinary Spatially Orientated Research...	132
	8.3 Regenerative Public Space in the City of Tshwane	134
	8.3.1 Phase 1: Common Understanding and Framework (2022)	135
	8.3.2 Phase 2: Understanding the Public Spaces and the Digital Platform.......................	136

		8.3.3 Phase 3: Testing and Launching the Digital Platform 140

      8.4 Blurring the Boundaries to Facilitate Deeper Levels of Collaboration, Participation and Cooperation 140

      8.5 Conclusion 142

      References.. 143

Conclusion: Recommendations for Spatial-Methodological, Transdisciplinary Action Regarding SDG11 (Position Paper).. 145

CONCLUSIÓN: Recomendaciones para una acción espacio-metodológica transdisciplinaria en relación con el ODS11 (Documento de posición) 151

Index.. 157

Introduction: Advancing Transdisciplinarity for Urban Sustainability Through Spatial Methods

Fraya Frehse ⓘ, Angela Million ⓘ, and Ignacio Castillo Ulloa ⓘ

Transdisciplinarity and Urban Sustainability

In the wake of close to a decade since the publication of the seventeenth United Nations (UN) Sustainable Development Goals (SDGs) in 2015, the sustainability scholarship throughout the world widely acknowledges that the accomplishment of sustainability in cities and communities, which lies at the core of SDG11, demands a *joint* knowledge production by academics and practitioners in local urban contexts (see among others Polk, 2015; Fam et al., 2017; Marshall et al., 2018; Sakao & Brambila-Macias, 2018; Kirby, 2019; Bojórquez-Tapia et al., 2021; Acksel, 2024). Underlying this approach is the concept of transdisciplinarity, which is conceptually grounded on the advocacy of research collaborations between academics and practitioners for the purpose of impacting the world of practice. Its furthest historic roots may be traced way back to the mid-twentieth century, whether in the problem-solving, action-research-based psychology bequeathed by the German-US American Kurt Lewin (1890–1947) (Hirsch Hadorn et al., 2008: 26) or in the problem-solving, participatory-research education developed by the Brazilian educator Paulo Freire (1921–1997) alongside the participatory action-research by the Colombian sociologist Orlando Fals Borda (1925–2008) (Brandão, 2005). This is not to mention the alleged first explicit use of the term 'transdisciplinarity' by the Swiss psychologist Jean Piaget (1896–1980) and a subsequent reflection about it by the Austrian-US American astrophysicist Erich Jantsch (1929–1980) at a seminar about interdisciplinarity in universities in 1970 (cf. respectively Piaget, 1972; Jantsch, 1970; for details see Bernstein, 2015: 2–7; for a counterpoint Jahn, 2012: 1, and Klein, 2014). Furthermore, according to Tanya Augsburg (2014) and Julie T. Klein (2004), there are two dominant transdisciplinary camps: (a) Nicolescuian transdisciplinarity (promoted by the Romanian physicist Basarab Nicolescu and the French philosopher Edgar Morin on the grounds that transdisciplinarity is a new methodology that allows to create new knowledge), and (b) the so-called Zurich School (originated after a conference held there in 2002, which conceptualises transdisciplinarity as a new kind of research, also called Mode 2 research, in which teams of researchers from diverse disciplines are brought together for brief periods of time to work intensively on specific problems to seek for solutions).

To be sure, the advocacy for transdisciplinarity in sustainability research, both within the urban and non-urban contexts, predates the publication of the UN Agenda 2030. Some authors argue that the 'discourse about transdisciplinarity' regarding social-ecological issues emerged in the forementioned 1950s (Miller et al., 2008: 9), 'between the climax of government-supported science and higher education and the long retrenchment that began in the 1970s' (Bernstein, 2015: 1). That said, the crucial aspect for us is that 'transdisciplinarity' has begun gaining traction during the 1990s in the framework of the historic emergence of truly 'global' debates on climate change and sustainability (Bernstein, 2015: 1). The plethora is immense (see for example Kates et al., 2001; Pohl et al., [2004] 2007; Hirsch Hadorn et al., 2008; Hunecke, 2011; Jahn et al., 2012; Brandt et al., 2013; Nicolescu, 2014). Given that these studies started to systematically tackle the complexity of these and other urban issues, which eventually came to be known as 'wicked problems' (Rittel & Weber, 1973; Brown et al., 2010), it became unescapable not to factor in methodological aspects. The overall scholarly interest in the 'how' of transdisciplinary collaborations in general (Bergmann et al., 2010; Dusseldorp & Beecroft, 2012; Popa et al., 2015; Bieluch et al., 2017; Craps, 2019; Fritz & Meinherz, 2020; Drake & Reid, 2021; Plummer et al., 2022; Philipp & Schmohl, 2023) soon started to co-exist with specific focuses on urban sustainability (Lang, 2012; Beland Lindahl & Westholm, 2014; Fam et al., 2018; von Wehrden et al., 2019; SMUS, 2020; Thondlhana et al., 2021; Sattler et al., 2022; Vienni-Baptista et al., 2023; Acksel, 2024).

The contribution of this book lies precisely in highlighting the potential of spatial methods at the intersection between transdisciplinarity and urban sustainability. While the range of methodological discussions regarding transdisciplinary projects for urban sustainability has been thriving vastly, little attention has been paid to *spatial* methods in particular. Therefore, this book looks into and, from diverse angles, sheds light on the following, twofold question: *Which possibilities and limitations can spatial methods respectively unravel and encounter for transdisciplinary research and practice, in view of the SDG11 targets?*

In doing so, four unique aspects come to the forefront: a focus on spatial methods in transdisciplinarity; an emphasis on case study projects located in the Global South; a use of a diverse range of methodologies; and an integration of both academic and practitioner (and even of citizens) perspectives within a collaborative framework.

Spatial Methods: Why?

Leaning on the Greek etymology of the word 'method' (*methodos*), the book concerns the various empirical research 'paths', or 'ways', which are available for scientific researchers, (local) practitioners (based, for instance, in NGOs, private firms or local government agencies), independent policy-makers and artists to co-produce *spatial* knowledge that may contribute to the attainment of the UN urban sustainability agenda. The adjective 'spatial' helps us heighten our specific methodological interest in this regard.

This book was conceived in the framework of our activities as Lead Partner and Action Speaker (Frehse), as Director and Action Speaker (Million), and as Scientific Coordinator and Action Speaker (Castillo Ulloa) of the Global Center of Spatial Methods for Urban Sustainability (SMUS). Founded in 2020, under the auspices of the German Academic Exchange Service (DAAD), the Center is the result of a broad international scientific and academic exchange project between Technische Universität (TU) Berlin and 47 universities from 7 regions across the so-called Global South, from Asia to Latin America to Africa (<htthttps://gcsmus.org>). The SMUS focus is on the transdisciplinary development and/or deployment of specific sets of research tools: from ethnographic observation to visualisation techniques (such as sketching, mental mapping, photographing and videos) as well as surveys and go-along interviews. As a common denominator, all these methodological devices enable a particular cognitive sensitivity, as it were, to the relations between human beings and material and symbolic goods alongside non-humans and more-than-human beings. In other words, the relations that make up 'space' as a socially, economically, politically, geographically and historically crucial category of thought and practice (for overviews see, among others, Lefebvre [1974] 2000; Löw, 2001; Dünne & Günzel, 2006; Günzel, 2009). Given this book's emphasis on *urban* sustainability, as addressed by SDG11, the space at stake comprises both 'cities and human settlements'. Hence, the book collaborators jointly assume these spaces as empirically referential for developing and/or deploying spatial methods in search of inclusion, safety, resilience and/or sustainability—which are the four basic elements underpinning SDG11's major objectives (UN, 2015).

While the role of spatial methods in spatial research within different academic disciplines has already been problematised (see for example Steinberg & Steinberg, 2012; Heinrich et al., 2024a, b; Rosenberg et al., 2022), the connection with transdisciplinary research and practice-oriented projects, aimed at implementing the UN urban sustainability agenda, remains visibly unexplored. On the other hand, there have already been transdisciplinary deployments of spatial methods (for example Hunecke, 2011; Lang, 2012; Beland Lindahl & Westholm, 2014; Fam et al., 2018; von Wehrden et al., 2019; Thondlhana et al., 2021; Sattler et al., 2022; Vienni-Baptista et al., 2023). But these studies lack critical assessments of the potential and challenges that the use of spatial methods for the purpose of the urban sustainability agenda entails (for exceptions see SMUS, 2020; Acksel, 2024).

Because this scholarly gap hints at a certain degree of thematic irrelevance, it begs the question: What could be the ultimate benefit of making spatial methods in transdisciplinarity for urban sustainability the central subject of a book?

Beyond its scholarly and editorial novelty, designed to provoke thought and practice, this issue leads to a core dilemma in transdisciplinary projects: How to effectively uphold low-impact urban design and development (LIUDD) (van Roon & van Roon, 2009; Lewis, 2009) through evidence-based and locally rooted knowledge (Davoudi, 2006, 2015; Faludi & Waterhoud, 2006). Finding ways to render this goal factually achievable requires the consideration of a wide range of micro-sociological, institutional and political challenging aspects that permeate the daily routine of transdisciplinary collaborations in both research and practice. Epistemologically, questions of 'how' are inherently methodological. And, in the case of transdisciplinarity for *urban* sustainability, answers to 'how' questions are bound to a spatial interface, given the essentially *spatial* nature of the 'cities and human settlements' addressed by SDG11. To sum up: from a methodological viewpoint, transdisciplinarity for urban sustainability has indeed a lot to profit from spatial methods.

Therefore, not only a whole new avenue (i.e. an alternative path-method!) come into scholarly reach, but also more adequate, faster and more effective paths for transdisciplinary collaboration towards urban sustainability are made scholarly available. This approach does not just open intellectual space for spatial-methodological exchanges between academics and practitioners. It also fosters joint critical reflections on the potential, drawbacks and inclusivity of spatial methods in transdisciplinary projects for the purpose of SDG11. Placed at the intersection of these intellectual possibilities and challenges, this book is one very particular outcome of the advocacy that spatial methods are essential, both as a new avenue and as a more inclusive and effective means, in transdisciplinarity for urban sustainability.

Spatial Methods: How?

This book's distinctiveness is not exclusively about addressing the development and/or deployment of spatial methods alongside their critical evaluation in transdisciplinary projects related to numerous SDG11 targets. Likewise, the authorship of the spatial-methodological engagement and assessment stands out. The next pages will present the reader with eight different transdisciplinary approaches to urban sustainability primarily based throughout the Global South by means of research-and-practice collaborations between scholars and practitioners. The spatial-methodological transdisciplinarity empirically took place in cities as diverse as Pretoria (South Africa), Porto Alegre and São Paulo (Brazil), Kolkata (India), Bangkok (Thailand) and Tshwane (South Africa) during the last 4 (partially Covid-19 pandemic) years, and in Tangerang (Indonesia) from 2001 to 2021. One individual, transdisciplinary trajectory in San José (Costa Rica) between 2004 and 2007 features a critical-reflexive methodological counterpoint.

This comprehensive globally south, geographic range of empirical references is especially suitable for addressing the book issue. The urban settings included in the chapters combine considerable empirical diversity regarding geographic, ecological as well as social-economic, social-cultural and historic aspects. At the same time, they all commonly share the remarkable co-existence of formal and informal urban-planning and regulatory initiatives by transdisciplinary stakeholders with very diverse social profiles, whether formally institutionalised or not, ranging from academics with varied disciplinary backgrounds to homeless people to schoolchildren, NGOs, CBOs and local communities. As the next pages make apparent, this combination of differences and similarities is methodologically especially helpful when it comes to addressing the core issue of the book. The Global South 'unity of differences' uniquely potentiates the scope of spatial-methodological strengths and weaknesses regarding transdisciplinarity for SDG11 that may be analytically depicted.

We now arrive at this book's fourth particularity. To enhance the analytical potential of the empirical cases even more, we gathered collaborators willing to pen their chapters transdisciplinarily as well. In the framework of critical assessments on the role of transdisciplinarity through spatial methods (Part I) or spatial methods through transdisciplinarity (Part II), all chapters are authored by academic-practitioner tandems (with the exception of the Chap. 2 author, who takes over the double role of academic-practitioner). The following pages will take the reader throughout the Global South by presenting perspectives and insights of: an urban planning professor, a knowledge applicator and a data strategist in Pretoria (Chap. 1); an architect formerly based in San José and engaged in (urban) master planning (Chap. 2); two professors of architecture and urban planning together with a freelance lawyer in Porto Alegre (Chap. 3); a professor and a doctoral student of sociology together with a public-policy project manager in São Paulo (Chap. 4); a professor and a research fellow of humanities and social sciences in conjunction with the leaseholder of a sewage-fed fish pond in Kolkata (Chap. 5); an anthropology professor, a master student alongside a marketer and naturalist in Bangkok (Chap. 6); a scholarly lecturer in environmental science and a water resource manager in Jakarta (Chap. 7); and a professor of town and regional planning and a professional town planner in Tshwane (Chap. 8).

In light of such a transdisciplinary diversity, it comes as no surprise that the definitions and understandings of both spatial methods and transdisciplinarity vary substantially across the chapters—not to mention their deployment. The methodological diversity coexists, and this is the book's fifth distinctiveness, with the prevalent use of two specific tools: mapping, whether or not within GIS (Chaps. 1, 2, 3, 7 and 8), and ethnography (Chaps. 4, 5 and 6).

Directly or indirectly, each chapter addresses the pros and cons of spatial methods for transdisciplinary projects regarding urban sustainability. In the conclusion, in turn, the majority of the authors get together to collectively outline a spatial-methodological, transdisciplinary roadmap. Indeed, the book ends with a set of four recommendations for spatial-methodological, transdisciplinary action for SDG11.

Given the vast combination of empirical, authorial and methodological elements that the book comprises, a second counterargument comes to the fore. Such a high degree of plurality would likely hinder any possibility of a synthetic answer to the matter at issue: bringing about urban sustainability by way of transdisciplinary and spatial research and practice.

Spatial Methods: Now What?

As legitimate as this criticism may well be, it also becomes somewhat inconsequential once it is considered that the book, by looking into the possibilities and limitations of spatial methods in transdisciplinarity for urban sustainability, puts forward one specific statement: spatial methods play a *transforma-*

tive role in transdisciplinary projects towards SDG11. In other words, their development, fine-tuning and/or deployment go hand in hand with more or less substantial *changes* for all the parties involved according to the local urban-spatial contexts in which the respective transdisciplinary research and/or practice-oriented projects take place.

It is thus no wonder that the transformative dimension of transdisciplinarity in sustainability constitutes, in and of itself, a research issue (see for example Fritz & Binder, 2018; Marshall et al., 2018). Transdisciplinary in both research and practice is by its very nature experimental, because the whole research-and/or-practice process is underpinned by a number of unavoidable contingencies—micro-sociological, cognitive, resource-related and communicative, among others. The following chapters well demonstrate that the combination of stakeholders involved in transdisciplinary research and practice-led projects for SDG11 is always circumstantial to some extent. Hence, transformation becomes both a throughput and output of the process, whether intended or not; and whether or not accompanied by the accomplishment of the explicitly pursued urban sustainability target. The *transformative* character endeavoured by transdisciplinary research and practice ideally takes shape, not just their actual implementation. It also underlies the delivered outcomes.

Indeed, more or less explicitly the case studies addressed in the book make evident that transdisciplinary spatial research and practice build on two distinctive modes of reflexivity: '*deliberative*' and '*pragmatist*' (Herrero et al., 2019: 752; *italics* in the original). Deliberative reflexivity attests to how much all actors involved weigh in with their argumentation on the values, epistemic basis and societal objectives to put into perspective both the practical matter at issue and its concomitant research question(s). In turn, pragmatist reflexivity refers to acquired and developed actor abilities, which emerge from equal-footing cooperative problem-solving and experimentation and allow the involved actors to (re)evaluate their comprehension of the practical problem situation as well as its derivative research question(s) (Ibid).

We are well aware of the existing and ongoing scholarly discussion on the issue; however, this book's approach to the matter is distinctive for it is spatial-methodological, as it were. The specific point we make concerns the transformative role of spatial methods in transdisciplinarity for urban sustainability—a dimension that arguably comes to the interpretive forefront of all the chapters. While only one addresses this aspect explicitly (Chap. 2), the other seven do it implicitly. The critical methodological reflections that make up each of the respective conclusive sections are enlightening in this regard. Transformation underpins post-fact assessments of respectively the varying author positionalities (Chap. 1), the conundrums of *practicing* transdisciplinarity concerning spatial methods employed as (urban) planning contrivances (Chap. 2), the disciplinary-boundary breaking role of spatial methods (Chaps. 3 and 4), the interactional role of communication via spatial methods (Chaps. 2, 4 and 6), their community-collaborative and resilient effects (Chaps. 5, 7 and 8).

With all this in mind, we invite the reader to delve into and engage in our spatial-methodological, transdisciplinary journey across the Global South for urban sustainability. Our hope is to reach an across-the-board disciplinary audience from the social and spatial sciences (sociology, anthropology, geography, architecture and urbanism), as well as practitioners, public servants and activists working on the target-subjects of UN SDG11.

All in all, this book aspires to deliver a substantive contribution to bridge the ever-lasting gap between academic and practice audiences in a unique way. It offers a conceptually and empirically based critical evaluation of the role of spatial methods in transdisciplinary research for the promotion of the UN Agenda 2030 concerning, specifically, the crisscrossing of urban inclusion, safety, resilience and sustainability.

Last but not least, to potentialise the global reach of this book, the introduction, conclusion, chapter titles, abstracts and keywords are also provided in Spanish.

Acknowledgements This book has been funded by the Global Center of Spatial Methods for Urban Sustainability (GCSMUS or SMUS), which is funded by the German Federal Ministry for Economic Cooperation and Development (BMZ) via the German Academic Exchange Service (DAAD)—Project No. 57526630.

References

Acksel, B. (2024). *Städte auf dem Weg zur Nachhaltigkeit*. Bielefeld: Transcript.

Augsburg, T. (2014). Becoming transdisciplinary: The emergence of the transdisciplinary individual. *The Journal of New Paradigm Research, 70*(3-4), 233–247.

Beland Lindahl, K., & Westholm, E. (2014). Transdisciplinarity in practice: Aims, collaboration and integration in a Swedish Research programme. *Journal of Integrative Environmental Sciences, 11*(3–4), 155–171. https://doi.org/10.1080/1943815X.2014.945940

Bergmann, M., Jahn, T., Knobloch, T., Krohn, W., Pohl, C., & Schramm, E. (Eds.). (2010). *Methoden transdisziplinärer Forschung*. Campus.

Bernstein, J. H. (2015). Transdisciplinarity: A review of its origins, development, and current issues. *Journal of Research Practice, 11*(1), Art. R1.

Bieluch, K. H., Bell, K. P., Teisl, M. F., Lindenfeld, L. A., Leahy, J., & Silka, L. (2017). Transdisciplinary research partnerships in sustainability science: An examination of stakeholder participation preferences. *Sustainability Science, 12*(1), 87–104. https://doi.org/10.1007/s11625-016-0360-x

Bojórquez-Tapia, L. A., Eakin, H., Hernández-Aguilar, B., & Shelton, R. (2021). Addressing complex, political and intransient sustainability challenges of transdisciplinarity: The case of the MEGADAP project in Mexico City. *Environmental Development, 38*(100604). https://doi.org/10.1016/j.envdev.2020.100604

Brandão, C. R. (2005). Participatory research and participation in research. A look between times and spaces from Latin America. *International Journal of Action Research, 1*(1), 43–688.

Brandt, P., Ernst, A., Gralla, F., et al. (2013). A review of transdiciplinary research in sustainability science. *Ecological Economy, 92*, 1–15.

Craps, M. (2019). In W. Leal Filho (Ed.), *Transdisciplinary processes for sustainable development*. Encyclopedia of Sustainability in Higher Education. https://doi.org/10.1007/978-3-319-63951-2_102-1

Davoudi, S. (2006). Evidence-based planning. *disP—The Planning Review, 165*(2), 14–25.

Davoudi, S. (2015). Planning as a practice of knowing. *Planning Theory, 14*(3), 316–331.

Drake, S., & Reid, J. (2021). Thinking now: Transdisciplinary thinking as a disposition. *Academia Letters, Art., 387*, 1–6. https://doi.org/10.20935/AL387.

Dünne, J., & Günzel, S. (Eds.). (2006). *Raumtheorie*. Frankfurt A. M.: Suhrkamp.

Dusseldorp, M., & Beecroft, R. (Eds.). (2012). *Technikfolgen abschätzen lehren*. Wiesbaden: Springer VS.

Faludi, A., & Waterhout, B. (2006). Introducing evidence-based planning. *disP—The Planning Review, 165*(2), 4–13.

Fam, D., Palmer, J., Riedy, C., & Mitchell, C. (Eds.). (2017). *Transdisciplinary Research and Practice for Sustainability Outcomes*. Routledge.

Fritz, L., & Meinherz, F. (2020). Tracing power in transdisciplinary sustainability research: An exploration. *Gaia, 29*(1), 41–51.

Günzel, S. (Ed.). (2009). *Raumwissenschaften*. Frankfurt a. M.

Heinrich, J.; Marguin, S.; Million, A.; Stollman. J. (eds.) (2024a). *Handbook of Qualitative and Visual Methods in Spatial Research*. Bielefeld: transcript.

Heinrich, J., Marguin, S., Million, A., & Stollman, J. (2024b). Spatial Research from an Interdisciplinary Perspective. In J. Heinrich, S. Marguin, A. Million, & J. Stollmann (Eds.), *Handbook of Qualitative and Visual Methods in Spatial Research* (pp. 9–16). (2021a). Bielefeld: Transcript.

Herrero, P., Dedeurwaerdere, T., & Osinki, A. (2019). Design features for social learning in transformative transdisciplinary research. *Sustainability Science, 14*, 751–769.

Hirsch Hadorn, G., Biber-Klemm, S., Grossenbacher-Mansuy, W., Hoffman-Riem, H., Joye, D., Pohl, C., Wiesmann, U., & Zemp, E. (2008a). The emergence of transdisciplinarity as a form of research. In G. Hirsch Hadorn, H. Hoffmann-Riem, S. Biber-Klemm, W. Grossenbacher-Mansuy, D. Joye, C. Pohl, U. Wiesmann, & E. Zemp (Eds.), *Handbook of Transdisciplinary Research* (pp. 19–39). Springer Science + Business Media B.V.

Hirsch Hadorn, G., Hoffmann-Riem, H., Biber-Klemm, S., Grossenbacher-Mansuy, W., Joye, D., Pohl, C., Wiesmann, U., & Zemp, E. (Eds.). (2008b). *Handbook of Transdisciplinary Research*. Springer Science + Business Media B.V.

Hunecke, M. (2011). Wissensintegration in der transdisziplinären Nachhaltigkeitsforschung: Eine Fallstudie zur Anpassung an zunehmende Starkniederschläge in urbanen Räumen. *Gaia, 20*(2), 104–111.

Jahn, T., Bergmann, M. and Keil, F. (2012). Transdisciplinarity: Between mainstreaming and marginalization. *Ecological Economics* 79: 1-10. https://doi. org/10.1016/j.ecolecon.2012.04.017.

Jantsch, E. (1970). Interdisciplinary and transdisciplinary university: Systems approach to educations and innovation. *Policy Sciences, 1*(4), 403–428.

Kates, R. W., Clark, W. C., & Corell, R. (2001). Environment and development—Sustainability science. *Science, 292*(5517), 641–642.

Kirby, A. (2019). Transdisciplinarity and sustainability science: A response to Sakao and Bambila-Macias in the context of sustainable cities research. *Journal of Cleaner Production, 210*, 238–245.

Klein, J. T. (2014). Discourses of transdisciplinarity: Looking back to the future. *Futures, 63*, 68–74.

Lam, D. P. M., Freund, M. E., Kny, J., Marg, O., Mbah, M., Theiler, L., Bergmann, M., Brohmann, B., Lang, D. J., & Schäfer, M. (2021). Transdisciplinary research: Towards an integrative perspective. *Gaia, 30*(4), 243–249.

Lawrence, R. (Ed.). (2023). *Handbook of Transdisciplinarity: Global Perspectives*. Edward Elgar Publishing.

Lefebvre, H. ([1974] 2000). *La production de l'espace*. Anthropos.

Lewis, L. (2009). The Integration of LID and Urban Design. In *Paper presented at the Flock Hill Workshop*. Urban Ecology and Ecological Design: Perspectives in Integration and Future Directions at Lincoln University.

Löw, M. (2001). *Raumsoziologie*. Frankfurt A. M.: Suhrkamp.

Marshall, F., Dolley, J., & Priya, R. (2018). Transdisciplinary research as transformative space making for sustainability: Enhancing propoor transformative agency in periurban contexts. *Ecology and Society, 23*(3), 8. https://doi.org/10.5751/ES-10249-230308

Mistra Urban Futures. (2021). *Transdisciplinary co-production as a tool for advancing SDGs implementation at the city level*. UN SDG Partnership Platform. https://sdgs.un.org/partnerships/transdisciplinary-co-production-tool-advancing-sdgs-implementation-city-level

Nicolescu, B. (2014). Multidisciplinarity, interdisciplinarity, indisciplinarity, and transdisciplinarity: Similarities and differences. In *RCC Perspectives 2* (pp. 19–26). Minding the Gap: Working Across Disciplines in Environmental Studies.

Philipp, T., & Schmohl, T. (Eds.). (2023). *Handbook Transdisciplinary Learning*. Transcript.

Piaget, J. (1972). The epistemology of interdisciplinary relationships. In Centre for Educational Research and Innovation (CERI). In *Interdisciplinarity: Problems of teaching and research in universities* (pp. 127–139). Organisation for Economic Co-operation and Development.

Plummer, R., Blythe, J., Gurney, G. G., Witkowski, S., & Armitage, D. (2022). Transdisciplinary partnerships for sustainability: An evaluation guide. *Sustainability Science, 17*, 955–967. https://doi.org/10.1007/s11625-021-01074-y

Pohl, C.; Hirsch Hadorn, G. ([2004] 2007). *Principles for designing transdisciplinary research. Proposed by the Swiss Academies of Arts and Sciences.* Oekom Verlag.

Polk, M. (2015). Transdiciplinary co-production: Designing and testing a transdisciplinary research framework for societal problem solving. *Futures, 65*, 110–122.

Popa, F., Guillermin, M., & Dedeurwaerdere, T. (2015). A pragmatist approach to transdisciplinarity in sustainability research: From complex systems theory to reflexive science. *Futures, 65*, 45–56. https://doi.org/10.1016/j.futures.2014.02.002

Renn, O. (2021). Transdisciplinarity: Synthesis towards a modular approach. *Futures, 130*(102744). https://doi.org/10.1016/j.futures.2021.102744

Rittel, H. W., & Webber, M. M. (1973). Dilemmas in a general theory of planning. *Policy Sciences, 4*, 155–169. https://doi.org/10.1007/BF01405730

Rosenberg, M. W., Lovell, S., & Coen, S. E. (Eds.). (2022). *The Routledge handbook of methodologies in human geography.* Routledge.

Sakao, T., & Brambila-Macias, S. A. (2018). Do we share an understanding of transdisciplinarity in environmental sustainability research'? *Journal of Cleaner Production, 170*, 1399–1403.

Sattler, C., Rommel, J., Chen, C., García-Llorente, M., Gutiérrez-Briceño, I., Prager, K., Reyes, M. F., Schröter, B., Schulze, C., van Bussel, L. G. J., Loft, L., Matzdorf, B., & Kelemen, E. (2022). Participatory research in times of COVID-19 and beyond: Adjusting your methodological toolkits. *One Earch, 5*, 62–65. https://doi.org/10.1016/j.oneear.2021.12.006

SMUS (Global Center of Spatial Methods for Urban Sustainability). (2020). *SMUS website.* https://gcsmus.org.

Steinberg, S. J., & Steinberg, S. L. (2012). *Geographic Information Systems for the Social Sciences: Investigating Space and Place.* Sage.

Thondhlana, G., Plaxedes Mubaya, C., McClure, A., Kwarteng Amaka-Otchere, A. B., & Ruwanza, S. (2021). Facilitating urban sustainability through transdiciplinary (TD) research: Lessons from Ghana, South Africa, and Zimbabwe. *Sustainability, 13*(6205), 1–18. https://doi.org/10.3390/su13116205

UN (United Nations). (2015). Make cities and human settlements inclusive, safe, resilient and sustainable. In: UN. Transforming our World: The 2030 Agenda for Sustainable Development. UN https://sdgs.un.org/goals/goal11.

van Roon, M. R., & van Roon, H. (2009). *Low impact urban design and development: The big picture.* Manaaki Whenua Press.

Vienni-Baptista, B., Fletcher, I., & Lyall, C. (Eds.). (2023). *Foundations of interdisciplinary and transdisciplinary research: A reader.* Bristol University Press.

von Wehrden, H., Guimarães, M. H., Bina, O., Varanda, M., Lang, D. J., John, B., Gralla, F., Alexander, D., Raines, D., White, A., & Lawrence, R. J. (2019). Interdisciplinary and transdisciplinary research: Finding the common ground of multi-faceted concepts. *Sustainability Science, 14*, 875–888. https://doi.org/10.1007/s11625-018-0594-x(012345

INTRODUCCIÓN: Transdisciplinariedad avanzada para la sostenibilidad urbana a través de métodos espaciales

Fraya Frehse ⓘ, Angela Million ⓘ, and Ignacio Castillo Ulloa ⓘ

Transdisciplinariedad y sostenibilidad urbana

Tras casi una década desde la publicación de los diecisiete Objetivos de Desarrollo Sostenible (ODS) de las Naciones Unidas (ONU) de 2015, los expertos en sostenibilidad de todo el mundo reconocen ampliamente que la consecución de la sostenibilidad en ciudades y comunidades –núcleo del ODS11— exige una producción conjunta de conocimientos por parte de académicos y profesionales en contextos urbanos locales (véanse entre otros Polk, 2015; Fam et al., 2017; Marshall et al., 2018; Sakao y Brambila-Macias, 2018; Kirby, 2019; Bojórquez-Tapia et al., 2021; Acksel, 2024). Este planteamiento es la base del concepto de transdisciplinariedad que, conceptualmente, se fundamenta en la defensa de las colaboraciones de investigación entre académicos y profesionales con el fin de influir en el mundo de la práctica. Las raíces históricas más lejanas se remontan a mediados del siglo XX, ya sea en la psicología basada en la resolución de problemas y la investigación de acción legada por el germano-estadounidense Kurt Lewin (1890–1947) (Hirsch Hadorn et al., 2008: 26) o en la educación basada en la resolución de problemas y la investigación participativa desarrollada por el educador brasileño Paulo Freire (1921–1997) además de la investigación de acción participativa del sociólogo colombiano Orlando Fals Borda (1925–2008) (Brandão, 2005). Esto sin mencionar el primer uso explícito del término «transdisciplinariedad» por el psicólogo suizo Jean Piaget (1896–1980) y la posterior reflexión del astrofísico austríaco-estadounidense Erich Jantsch (1929–1980) en un seminario sobre la interdisciplinaridad en las universidades en 1970 (cf. respectivamente Piaget, 1972; Jantsch, 1970; para más detalles, véase Bernstein, 2015: 2–7 y para un contrapunto a Jahn, 2012: 1, y Klein, 2014). Además, según Tanya Augsburg (2014) y Julie T. Klein (2004) existen dos campos predominantes en la transdisciplinariedad: (a) la transdisciplinariedad de Nicolescu (promovida por el físico rumano Basarab Nicolescu y el filósofo francés Edgar Morin considerando que la transdisciplinariedad es una nueva metodología que permite crear nuevos conocimientos) y (b) la denominada Escuela de Zúrich (que surgió después de una conferencia celebrada allí en 2002 y que conceptualiza la transdisciplinariedad como un nuevo tipo de investigación—también denominada investigación de modo 2—en el que equipos de investigadores de varias disciplinas se reúnen durante breves periodos de tiempo para trabajar intensamente en problemas específicos para buscar soluciones).

Sin duda, la defensa de la transdisciplinariedad en la investigación de la sostenibilidad, tanto en el contexto urbano como en el no urbano, es previa a la publicación de la Agenda 2030 de la ONU. Algunos autores argumentan que el «discurso sobre la transdisciplinariedad» en relación con aspectos socioecológicos surgió en la mencionada década de 1950 (Miller et al., 2008: 9) «entre el clímax de la ciencia y la enseñanza superior apoyadas por el gobierno y los grandes recortes que empezaron en los años setenta» (Bernstein, 2015: 1).Dicho esto, el aspecto crucial que nos incumbe es que la «transdisciplinariedad» empezó a ganar terreno durante los años noventa en el marco de la emergencia histórica de debates verdaderamente «globales» sobre el cambio climático y la sostenibilidad (Bernstein, 2015: 1). La plétora es inmensa (véase, por ejemplo, Kates et al., 2001; Pohl et al., [2004] 2007; Hirsch Hadorn et al., 2008; Hunecke, 2011; Jahn et al., 2012; Brandt et al., 2013; Nicolescu, 2014). Dado que estos estudios empezaron a abordar de manera sistemática la complejidad de estos y otros aspectos urbanísticos, que acabaron conociéndose como «problemas retorcidos» (Rittel y Weber 1973; Brown et al., 2010), resultaba ineludible no tener en cuenta los aspectos metodológicos.El interés académico global por el «cómo» de las colaboraciones transdisciplinarias en general (Bergmann et al., 2010; Dusseldorp y Beecroft, 2012; Popa et al., 2015; Bieluch et al., 2017; Craps, 2019; Fritz y Meinherz, 2020; Drake y Reid, 2021; Plummer et al., 2022; Philipp y Schmohl, 2023) empezó pronto a coexistir con enfoques específicos sobre la sostenibilidad urbana (Lang, 2012; Beland Lindahl y Westholm, 2014; Fam et al., 2018; von Wehrden et al., 2019; SMUS, 2020; Thondlhana et al., 2021; Sattler et al., 2022; Vienni-Baptista et al., 2023; Acksel, 2024).

La contribución del presente libro radica precisamente en destacar el potencial de los métodos espaciales en la intersección entre transdisciplinariedad y sostenibilidad urbana. Aunque la variedad de debates metodológicos relacionados con proyectos transdisciplinarios para la sostenibilidad urbana ha ido aumentando considerablemente, se ha prestado una escasa atención a los métodos espaciales en particular. Por eso, este libro examina y arroja luz desde diversos ángulos a la siguiente doble cuestión: ¿qué posibilidades y limitaciones pueden desentrañar y encontrar, respectivamente, los métodos espaciales para la investigación y la práctica transdisciplinarias de cara a los objetivos del ODS11?

Hay cuatro aspectos únicos que destacan: la atención a los métodos espaciales en la transdisciplinariedad; el énfasis sobre proyectos de estudios de casos localizados en el sur global; el uso de una amplia variedad de metodologías; y la integración de las perspectivas académicas y profesionales (e incluso de los ciudadanos) dentro de un marco de colaboración.

Métodos espaciales: ¿por qué?

Apoyándose en la etimología griega de la palabra «método» (methodos), el libro se ocupa de varias «vías» o «caminos» de investigación empírica a disposición de los investigadores científicos, profesionales (locales) (por ejemplo, ONG, empresas privadas o agencias gubernamentales locales), responsables políticos y artistas independientes para coproducir conocimientos espaciales que pueda contribuir a la consecución de la agenda de sostenibilidad urbana de la ONU. El adjetivo «espacial» nos ayuda a reforzar nuestro interés metodológico específico sobre esta cuestión.

Este libro ha sido concebido en el marco de nuestras actividades como socio líder y ponente de acción (Frehse), como director y ponente de acción (Million) y como coordinador científico y ponente de acción (Castillo Ulloa) del Centro Global de Métodos Espaciales para la Sostenibilidad Urbana (SMUS). Este centro, fundado en 2020, bajo la supervisión del Servicio Alemán de Intercambio Académico (DAAD), es el resultado de un amplio proyecto de intercambio científico y académico internacional entre la Universidad Técnica (TU) de Berlín y 47 universidades de 7 regiones del denominado sur global, que engloba desde Asia hasta América Latina y África (<htthttps://gcsmus.org>). El SMUS se centra en el desarrollo y/o despliegue transdisciplinarios de conjuntos específicos de herramientas de investigación: desde la observación etnográfica hasta las técnicas de visualización (como bocetos, mapas mentales, fotografías y vídeos), así como encuestas y entrevistas. Como denominador común, todos estos dispositivos metodológicos permiten, por así decirlo, una sensibilidad cognitiva particular a las relaciones entre los seres humanos y los bienes materiales y simbólicos junto con los no humanos y sobrehumanos. Dicho de otro modo, las relaciones que conforman el «espacio» como categoría de pensamiento y práctica social, económica, política, geográfica e históricamente crucial (para una visión general, véanse, entre otros, Lefebvre, [1974] 2000; Löw, 2001; Dünne y Günzel, 2006; Günzel, 2009). Dado el énfasis que da este libro a la sostenibilidad urbana, tal y como se aborda en el ODS11, el espacio en cuestión comprende tanto las «ciudades como los asentamientos humanos». De ahí que los colaboradores del libro asuman conjuntamente estos espacios como referentes empíricos para desarrollar y/o desplegar métodos espaciales en busca de la inclusión, la seguridad, la resiliencia y/o la sostenibilidad, que son los cuatro pilares del objetivo principal del ODS11 (ONU, 2015).

Aunque ya se ha problematizado la función de los métodos espaciales en la investigación espacial dentro de las diferentes disciplinas académicas (véase, por ejemplo, Steinberg y Steinberg, 2012; Heinrich et al., 2024a, b; Rosenberg et al., 2022), la conexión con la investigación transdisciplinaria y los proyectos orientados a la práctica, dirigidos a implementar la agenda de sostenibilidad urbana de la ONU, sigue estando visiblemente sin explorar. Por otro lado, ya ha habido despliegues transdisciplinarios de métodos espaciales (por ejemplo, Hunecke, 2011; Lang, 2012; Beland Lindahl y Westholm, 2014; Fam et al., 2018; von Wehrden et al., 2019; Thondlhana et al., 2021; Sattler et al., 2022; Vienni-Baptista et al., 2023). No obstante, dichos estudios carecen de evaluaciones críticas sobre el potencial y los desafíos que conlleva el uso de métodos espaciales para el propósito de la agenda de sostenibilidad urbana (para excepciones, véase SMUS, 2020; Acksel, 2024).

Dado que esta laguna académica deja entrever un cierto grado de irrelevancia temática, cabe preguntarse lo siguiente: ¿cuál podría ser el beneficio final de centrar el libro en los métodos espaciales en la transdisciplinariedad para la sostenibilidad urbana?

Más allá de la novedad académica y editorial –que trata de incitar a la reflexión y a la práctica –, esta cuestión conduce al dilema central de los proyectos transdisciplinarios: cómo defender eficazmente el diseño y desarrollo urbano de bajo impacto (LIUDD, por su sigla en inglés) (van Roon y van Roon, 2009; Lewis, 2009) a través de conocimientos basados en pruebas y arraigados localmente (Davoudi, 2006, 2015; Faludi y Waterhoud, 2006). Encontrar la manera de hacer factible este objetivo requiere considerar un amplio abanico de aspectos microsociológicos, institucionales y políticos desafiantes que impregnan la rutina diaria de las colaboraciones transdisciplinarias tanto en la investigación como en la práctica. Epistemológicamente, las cuestiones sobre el «cómo» son intrínsecamente metodológicas. En el caso particular de la transdisciplinariedad para la sostenibilidad urbana, las respuestas a las preguntas sobre el «cómo» están vinculadas a una interfaz espacial dada la naturaleza esencialmente espacial de las «ciudades y asentamientos humanos» que se abordan en el ODS11. En resumen, desde un punto de vista metodológico, la transdisciplinariedad para la sostenibilidad urbana tiene mucho de lo que beneficiarse de los métodos espaciales.

Por lo tanto, no solo se abre una vía totalmente nueva (por ejemplo, un método alternativo) para los académicos, sino que también se ponen a su disposición vías más adecuadas, más rápidas y más eficaces para la colaboración transdisciplinaria en pro de la sostenibilidad urbana. Este planteamiento no solo abre un espacio intelectual para el intercambio de metodologías espaciales entre académicos y profesionales. También fomenta las reflexiones críticas conjuntas sobre el potencial, los inconvenientes y la inclusividad de los métodos espaciales en proyectos transdisciplinarios para el propósito del ODS11. Este libro, situado en la intersección intelectual de estas posibilidades y retos, es un resultado muy particular de la defensa de que los métodos espaciales son esenciales –como una nueva vía y como medio más inclusivo y eficaz– en la transdisciplinariedad para sostenibilidad urbana.

Métodos espaciales: ¿cómo?

El carácter distintivo de este libro no tiene que ver exclusivamente con el hecho de abordar el desarrollo y/o despliegue de métodos espaciales junto con su evaluación crítica en proyectos transdisciplinarios relacionados con varios objetivos del ODS11. Asimismo, destaca la autoría del compromiso y la evaluación de la metodología espacial. En las siguientes páginas se presentan al lector ocho enfoques transdisciplinarios distintos de la sostenibilidad urbana basados principalmente en todo el sur global por medio de colaboraciones de investigación y práctica entre académicos y profesionales. La transdisciplinariedad metodológica espacial tuvo lugar, empíricamente, en ciudades tan diversas como Pretoria (Sudáfrica), Porto Alegre y São Paulo (Brasil), Calcuta (India), Bangkok (Tailandia) y Tshwane (Sudáfrica) durante los últimos cuatro años (parcialmente durante la pandemia de la COVID-19) y en Tangerang (Indonesia) de 2001 a 2021. Una trayectoria individual y transdisciplinaria en San José (Costa Rica) entre 2004 y 2007 presenta un contrapunto metodológico crítico y reflexivo.

Esta amplia zona geográfica (sur global) de referencias empíricas es especialmente adecuada para abordar la cuestión de este libro. Los entornos urbanos que se incluyen en los capítulos combinan una considerable diversidad empírica con respecto a aspectos geográficos, ecológicos y socioeconómicos, socioculturales e históricos. Al mismo tiempo, todos tienen en común la remarcable coexistencia de iniciativas formales e informales de planificación y regulación urbana por parte de interesados transdisciplinarios con perfiles sociales muy diversos, formalmente institucionalizados o no, que van desde académicos de diferentes ámbitos profesionales hasta personas sin hogar y escolares, ONG, organizaciones basadas en la comunidad y comunidades locales. Como se pone de manifiesto en las siguientes páginas, esta combinación de diferencias y similitudes resulta especialmente práctica desde el punto de vista metodológico cuando se trata de abordar el tema principal de este libro. La «unidad de las diferencias» del sur global potencia de forma única el alcance de las fortalezas y debilidades de la metodología espacial relativas a la transdisciplinariedad para el ODS11 que pueden representarse analíticamente.

Ahora, vamos a abordar la cuarta particularidad de este libro. Para aumentar aún más el potencial analítico de los casos empíricos, reunimos a colaboradores dispuestos a redactar sus capítulos también transdisciplinariamente. En el marco de las evaluaciones críticas sobre el papel de la transdisciplinariedad a través de métodos espaciales (Parte I) o los métodos espaciales a través de la transdisciplinariedad (Parte II), todos los capítulos están redactados por tándems de académicos y profesionales (a excepción del capítulo 2, cuyo autor es académico y profesional a la vez). Las siguientes páginas harán viajar al lector por el globo meridional presentando las perspectivas y percepciones de: un profesor de urbanismo, un aplicador de conocimientos y estratega de datos de Pretoria (capítulo 1); un arquitecto afincado en San José y dedicado a la planificación general (urbana) (capítulo 2); dos profesores de arquitectura y urbanismo junto con un abogado autónomo en Porto Alegre (capítulo 3); un profesor y un doctorando en sociología junto con un gestor de proyectos de políticas públicas en São Paulo (capítulo 4); un profesor y un becario de investigación de humanidades y ciencias sociales junto con el arrendatario de un estanque de peces alimentado por aguas residuales en Calcuta (capítulo 5); un profesor de antropología, un estudiante de máster junto a un comercial y naturalista en Bangkok (capítulo 6); un profesor de ciencias ambientales y gestión de recursos hídricos en Yakarta (capítulo 7); un profesor de planificación urbana y regional y un profesional en urbanismo en Tshwane (capítulo 8).

Ante tal diversidad transdisciplinaria, no es de extrañar que las definiciones y entendimientos de los métodos espaciales y de la transdisciplinariedad varíen sustancialmente de un capítulo a otro, por no mencionar su despliegue. La diversidad metodológica coexiste –y este es el quinto rasgo distintivo del libro– con el uso predominante de dos herramientas especificas: el mapeo, dentro o fuera del GIS, (capítulos 1, 2, 3, 7 y 8) y la etnografía (capítulos 4, 5 y 6).

Directa o indirectamente, cada capítulo aborda los pros y contras de los métodos espaciales para proyectos transdisciplinarios relacionados con la sostenibilidad urbana. En la conclusión, en cambio, la mayoría de los autores se unen para esbozar colectivamente una hoja de ruta metodológica-espacial y transdisciplinaria. De hecho, el libro termina con un conjunto de cuatro recomendaciones para la acción transdisciplinaria y de metodológica-espacial para el ODS11.

Dada la vasta combinación de elementos empíricos, metodológicos y de autores que aborda el libro, surge un segundo contraargumento. Un grado tan elevado de pluralidad probablemente obstaculizaría cualquier posibilidad de dar una respuesta sintética al problema: lograr la sostenibilidad urbana por medio de la investigación y la práctica transdisciplinaria y espacial.

Métodos espaciales: ¿ahora qué?

Por muy legítima que pueda ser esta crítica, también se convierte en algo intrascendente una vez que se considera que el libro, al examinar las posibilidades y limitaciones de los métodos espaciales en la transdisciplinariedad para la sostenibilidad urbana, expone una afirmación específica: los métodos espaciales desempeñan un papel transformador en los proyectos transdisciplinarios en pro del ODS11. En otras palabras, su desarrollo, perfeccionamiento y/o despliegue van de la mano de cambios más o menos sustanciales para todas las partes involucradas según los contextos urbano-espaciales locales en los que tienen lugar los respectivos proyectos transdisciplinarios de investigación y/u orientados a la práctica.

Así pues, no sorprende que la dimensión transformadora de la transdisciplinariedad en cuestiones de sostenibilidad sea, en sí misma, un tema de investigación (véanse, por ejemplo, Fritz y Binder, 2018; Marshall et al., 2018). La transdisciplinariedad en la investigación y en la práctica es, por su propia naturaleza, experimental, ya que todo el proceso de investigación y/o práctica se sustenta en una serie de contingencias inevitables: microsociológicas, cognitivas, relacionadas con los recursos y comunicativas, entre otras. Los siguientes capítulos demuestran bien que la combinación de las partes implicadas en los proyectos transdisciplinarios de investigación y práctica para el ODS11 es siempre circunstancial hasta cierto punto. Por lo tanto, la transformación se convierte en un producto y en un resultado del proceso, ya sea intencionado o no y vaya acompañado o no de la consecución del objetivo de sostenibilidad urbana que se persigue explícitamente. Lo ideal es que el carácter transformador que persiguen la investigación y la práctica transdisciplinarias no solo se plasme en su aplicación real, sino que también se sustente en los resultados obtenidos.

De hecho, de forma más o menos explícita, los estudios de caso abordados en este libro ponen de manifiesto que la investigación y la práctica transdisciplinarias espaciales se basan en dos modos distintos de reflexividad: la «deliberativa» y la «pragmatista» (Herrero et al., 2019: 752; en cursiva en el original). La reflexividad deliberativa acredita hasta qué punto todos los actores implicados aportan su argumentación sobre los valores, la base epistémica y los objetivos sociales para poner en perspectiva tanto la cuestión práctica del problema como su(s) cuestión(es) de investigación concomitante(s). A su vez, la reflexividad pragmatista se refiere a las habilidades adquiridas y desarrolladas por los actores, que emergen de la resolución cooperativa de problemas y de la experimentación en igualdad de condiciones, y permiten a las partes involucradas (re)evaluar su comprensión sobre la situación práctica del problema y la(s) cuestión(es) de investigación derivada(s) (Ibid).

Somos muy conscientes del debate académico existente y presente sobre esta cuestión. Sin embargo, el planteamiento de este libro sobre esta materia se distingue por su metodología espacial, por así decirlo. La cuestión especifica que planteamos se refiere al papel transformador de los métodos espaciales en la transdisciplinariedad para la sostenibilidad urbana, una dimensión que posiblemente pase al primer plano interpretativo de todos los capítulos. Aunque solo uno de los capítulos aborda este aspecto explícitamente (capítulo 2), los otros siete lo hacen implícitamente. Las reflexiones metodológicas críticas que componen cada una de las respectivas secciones concluyentes son esclarecedoras a este respecto. La transformación sustenta las evaluaciones posteriores a los hechos de, respectivamente, las diferentes posturas de los autores (capítulo 1), los entresijos de la práctica transdisciplinaria en relación con los métodos espaciales empleados como artificios de planificación (urbana) (capítulo 2), el papel de ruptura de fronteras disciplinarias de los métodos espaciales (capítulos 3 y 4), el papel interactivo de la comunicación a través de los métodos espaciales (capítulos 2, 4 y 6) y sus efectos de colaboración comunitaria y de resiliencia (capítulos 5, 7 y 8).

Con todo esto presente, invitamos al lector a adentrarse y participar en nuestro viaje metodológico-espacial y transdisciplinar por el sur global en pos de la sostenibilidad urbana. Esperamos poder llegar a un público que abarque todas las disciplinas de las ciencias sociales y espaciales (sociología, antropología, geografía, arquitectura y urbanismo), así como a profesionales, funcionarios públicos y activistas que trabajen en las cuestiones relacionadas con el ODS11 de la ONU.

En conjunto, este libro aspira a una contribución sustancial para tender puentes entre el público académico y el práctico de una forma única. Además, ofrece una evaluación crítica con base conceptual y empírica del papel de los métodos espaciales de la investigación transdisciplinaria para promover la Agenda 2030 de la ONU en lo que respecta, específicamente, al cruce de la inclusión, la seguridad, la resiliencia y la sostenibilidad urbanas.

Por último, pero no por ello menos importante, para potenciar el alcance mundial de este libro, la introducción, la conclusión, los títulos de los capítulos, los resúmenes y las palabras clave se ofrecen también en español.

Reconocimientos Este libro ha sido financiado por el Centro Global de Métodos Espaciales para la Sostenibilidad Urbana (GCSMUS o SMUS), financiado a su vez por el Ministerio Federal Alemán de Cooperación Económica y Desarrollo (BMZ) a través del Servicio Alemán de Intercambio Académico (DAAD) – N.° de proyecto 57526630.

Referencias

Acksel, B. (2024). *Städte auf dem Weg zur Nachhaltigkeit*. Bielefeld: Transcript.

Augsburg, T. (2014). Becoming transdisciplinary: The emergence of the transdisciplinary individual. *The Journal of New Paradigm Research, 70*(3-4), 233–247.

Beland Lindahl, K., & Westholm, E. (2014). Transdisciplinarity in practice: Aims, collaboration and integration in a Swedish Research programme. *Journal of Integrative Environmental Sciences, 11*(3–4), 155–171. https://doi.org/10.1080/1943815X.2014.945940

Bergmann, M., Jahn, T., Knobloch, T., Krohn, W., Pohl, C., & Schramm, E. (Eds.). (2010). *Methoden transdisziplinärer Forschung*. Campus.

Bernstein, J. H. (2015). Transdisciplinarity: A review of its origins, development, and current issues. *Journal of Research Practice, 11*(1), Art. R1.

Bieluch, K. H., Bell, K. P., Teisl, M. F., Lindenfeld, L. A., Leahy, J., & Silka, L. (2017). Transdisciplinary research partnerships in sustainability science: An examination of stakeholder participation preferences. *Sustainability Science, 12*(1), 87–104. https://doi.org/10.1007/s11625-016-0360-x

Bojórquez-Tapia, L. A., Eakin, H., Hernández-Aguilar, B., & Shelton, R. (2021). Addressing complex, political and intransient sustainability challenges of transdisciplinarity: The case of the MEGADAP project in Mexico City. *Environmental Development, 38*(100604). https://doi.org/10.1016/j.envdev.2020.100604

Brandão, C. R. (2005). Participatory research and participation in research. A look between times and spaces from Latin America. *International Journal of Action Research, 1*(1), 43–688.

Brandt, P., Ernst, A., Gralla, F., et al. (2013). A review of transdisciplinary research in sustainability science. *Ecological Economy, 92*, 1–15.

Craps, M. (2019). In W. Leal Filho (Ed.), *Transdisciplinary processes for sustainable development*. Encyclopedia of Sustainability in Higher Education. https://doi.org/10.1007/978-3-319-63951-2_102-1

Davoudi, S. (2006). Evidence-based planning. *disP—The Planning Review, 165*(2), 14–25.

Davoudi, S. (2015). Planning as a practice of knowing. *Planning Theory, 14*(3), 316–331.

Drake, S., & Reid, J. (2021). Thinking now: Transdisciplinary thinking as a disposition. *Academia Letters, Art., 387*, 1–6. https://doi.org/10.20935/AL387.

Dünne, J., & Günzel, S. (Eds.). (2006). *Raumtheorie*. Frankfurt A. M.: Suhrkamp.

Dusseldorp, M., & Beecroft, R. (Eds.). (2012). *Technikfolgen abschätzen lehren*. Wiesbaden: Springer VS.

Faludi, A., & Waterhout, B. (2006). Introducing evidence-based planning. *disP—The Planning Review, 165*(2), 4–13.

Fam, D., Palmer, J., Riedy, C., & Mitchell, C. (Eds.). (2017). *Transdisciplinary Research and Practice for Sustainability Outcomes*. Routledge.

Fritz, L., & Meinherz, F. (2020). Tracing power in transdisciplinary sustainability research: An exploration. *Gaia, 29*(1), 41–51.

Günzel, S. (Ed.). (2009). *Raumwissenschaften*. Frankfurt a. M.

Heinrich, J.; Marguin, S.; Million, A.; Stollman. J. (eds.) (2024a). *Handbook of Qualitative and Visual Methods in Spatial Research*. Bielefeld: transcript.

Heinrich, J., Marguin, S., Million, A., & Stollman, J. (2024b). Spatial Research from an Interdisciplinary Perspective. In J. Heinrich, S. Marguin, A. Million, & J. Stollmann (Eds.), *Handbook of Qualitative and Visual Methods in Spatial Research* (pp. 9–16). (2021a). Bielefeld: Transcript.

Herrero, P., Dedeurwaerdere, T., & Osinki, A. (2019). Design features for social learning in transformative transdisciplinary research. *Sustainability Science, 14*, 751–769.

Hirsch Hadorn, G., Biber-Klemm, S., Grossenbacher-Mansuy, W., Hoffman-Riem, H., Joye, D., Pohl, C., Wiesmann, U., & Zemp, E. (2008a). The emergence of transdisciplinarity as a form of research. In G. Hirsch Hadorn, H. Hoffmann-Riem, S. Biber-Klemm, W. Grossenbacher-Mansuy, D. Joye, C. Pohl, U. Wiesmann, & E. Zemp (Eds.), *Handbook of Transdisciplinary Research* (pp. 19–39). Springer Science + Business Media B.V.

Hirsch Hadorn, G., Hoffmann-Riem, H., Biber-Klemm, S., Grossenbacher-Mansuy, W., Joye, D., Pohl, C., Wiesmann, U., & Zemp, E. (Eds.). (2008b). *Handbook of Transdisciplinary Research*. Springer Science + Business Media B.V.

Hunecke, M. (2011). Wissensintegration in der transdisziplinären Nachhaltigkeitsforschung: Eine Fallstudie zur Anpassung an zunehmende Starkniederschläge in urbanen Räumen. *Gaia, 20*(2), 104–111.

Jahn, T., Bergmann, M. and Keil, F. (2012). Transdisciplinarity: Between mainstreaming and marginalization. *Ecological Economics* 79: 1-10. https://doi. org/10.1016/j.ecolecon.2012.04.017.

Jantsch, E. (1970). Interdisciplinary and transdisciplinary university: Systems approach to educations and innovation. *Policy Sciences, 1*(4), 403–428.

Kates, R. W., Clark, W. C., & Corell, R. (2001). Environment and development—Sustainability science. *Science, 292*(5517), 641–642.

Kirby, A. (2019). Transdisciplinarity and sustainability science: A response to Sakao and Bambila-Macias in the context of sustainable cities research. *Journal of Cleaner Production, 210*, 238–245.

Klein, J. T. (2014). Discourses of transdisciplinarity: Looking back to the future. *Futures, 63*, 68–74.

Lam, D. P. M., Freund, M. E., Kny, J., Marg, O., Mbah, M., Theiler, L., Bergmann, M., Brohmann, B., Lang, D. J., & Schäfer, M. (2021). Transdisciplinary research: Towards an integrative perspective. *Gaia, 30*(4), 243–249.

Lawrence, R. (Ed.). (2023). *Handbook of Transdisciplinarity: Global Perspectives*. Edward Elgar Publishing.

Lefebvre, H. ([1974] 2000). *La production de l'espace*. Anthropos.

Lewis, L. (2009). *The Integration of LID and Urban Design*. In *Paper presented at the Flock Hill Workshop*. Urban Ecology and Ecological Design: Perspectives in Integration and Future Directions at Lincoln University.

Löw, M. (2001). *Raumsoziologie*. Frankfurt A. M.: Suhrkamp.

Marshall, F., Dolley, J., & Priya, R. (2018). Transdisciplinary research as transformative space making for sustainability: Enhancing propoor transformative agency in periurban contexts. *Ecology and Society, 23*(3), 8. https://doi.org/10.5751/ES-10249-230308

Mistra Urban Futures. (2021). *Transdisciplinary co-production as a tool for advancing SDGs implementation at the city level*. UN SDG Partnership Platform. https://sdgs.un.org/partnerships/transdisciplinary-co-production-tool-advancing-sdgs-implementation-city-level

Nicolescu, B. (2014). Multidisciplinarity, interdisciplinarity, indisciplinarity, and transdisciplinarity: Similarities and differences. In *RCC Perspectives 2* (pp. 19–26). Minding the Gap: Working Across Disciplines in Environmental Studies.

Philipp, T., & Schmohl, T. (Eds.). (2023). *Handbook Transdisciplinary Learning*. Transcript.

Piaget, J. (1972). The epistemology of interdisciplinary relationships. In Centre for Educational Research and Innovation (CERI). In *Interdisciplinarity: Problems of teaching and research in universities* (pp. 127–139). Organisation for Economic Co-operation and Development.

Plummer, R., Blythe, J., Gurney, G. G., Witkowski, S., & Armitage, D. (2022). Transdisciplinary partnerships for sustainability: An evaluation guide. *Sustainability Science, 17*, 955–967. https://doi.org/10.1007/s11625-021-01074-y

Pohl, C.; Hirsch Hadorn, G. ([2004] 2007). *Principles for designing transdisciplinary research. Proposed by the Swiss Academies of Arts and Sciences*. Oekom Verlag.

Polk, M. (2015). Transdiciplinary co-production: Designing and testing a transdisciplinary research framework for societal problem solving. *Futures, 65*, 110–122.

Popa, F., Guillermin, M., & Dedeurwaerdere, T. (2015). A pragmatist approach to transdisciplinarity in sustainability research: From complex systems theory to reflexive science. *Futures, 65*, 45–56. https://doi.org/10.1016/j.futures.2014.02.002

Renn, O. (2021). Transdisciplinarity: Synthesis towards a modular approach. *Futures, 130*(102744). https://doi.org/10.1016/j.futures.2021.102744

Rittel, H. W., & Webber, M. M. (1973). Dilemmas in a general theory of planning. *Policy Sciences, 4*, 155–169. https://doi.org/10.1007/BF01405730

Rosenberg, M. W., Lovell, S., & Coen, S. E. (Eds.). (2022). *The Routledge handbook of methodologies in human geography*. Routledge.

Sakao, T., & Brambila-Macias, S. A. (2018). Do we share an understanding of transdisciplinarity in environmental sustainability research'? *Journal of Cleaner Production, 170*, 1399–1403.

Sattler, C., Rommel, J., Chen, C., García-Llorente, M., Gutiérrez-Briceño, I., Prager, K., Reyes, M. F., Schröter, B., Schulze, C., van Bussel, L. G. J.,

Loft, L., Matzdorf, B., & Kelemen, E. (2022). Participatory research in times of COVID-19 and beyond: Adjusting your methodological toolkits. *One Earth, 5*, 62–65. https://doi.org/10.1016/j.oneear.2021.12.006

SMUS (Global Center of Spatial Methods for Urban Sustainability). (2020). *SMUS website*. https://gcsmus.org.

Steinberg, S. J., & Steinberg, S. L. (2012). *Geographic Information Systems for the Social Sciences: Investigating Space and Place*. Sage.

Thondhlana, G., Plaxedes Mubaya, C., McClure, A., Kwarteng Amaka-Otchere, A. B., & Ruwanza, S. (2021). Facilitating urban sustainability through transdiciplinary (TD) research: Lessons from Ghana, South Africa, and Zimbabwe. *Sustainability, 13*(6205), 1–18. https://doi.org/10.3390/su13116205

UN (United Nations). (2015). Make cities and human settlements inclusive, safe, resilient and sustainable. In: UN. Transforming our World: The 2030 Agenda for Sustainable Development. UN https://sdgs.un.org/goals/goal11.

van Roon, M. R., & van Roon, H. (2009). *Low impact urban design and development: The big picture*. Manaaki Whenua Press.

Vienni-Baptista, B., Fletcher, I., & Lyall, C. (Eds.). (2023). *Foundations of interdisciplinary and transdisciplinary research: A reader*. Bristol University Press.

von Wehrden, H., Guimarães, M. H., Bina, O., Varanda, M., Lang, D. J., John, B., Gralla, F., Alexander, D., Raines, D., White, A., & Lawrence, R. J. (2019). Interdisciplinary and transdisciplinary research: Finding the common ground of multi-faceted concepts. *Sustainability Science, 14*, 875–888. https://doi.org/10.1007/s11625-018-0594-x(012345

Part I
The Critical Role of Transdisciplinarity Through Spatial Methods

A Methodological Framework for Transdisciplinary Urban Planning

Jacques du Toit, Amy Pieterse, and Sandile Mbatha

1.1 Introduction

Urban planning is an applied social science with a strong normative orientation to improve the human condition through the restorative, equitable and sustainable management and development of land. One of the broader trends in research since the 1990s, including the applied social sciences, has been a shift from Mode 1 knowledge production, the positivist conception of a singular body of knowledge produced by researchers and feeding neatly into practice, to Mode 2 knowledge production, the realist conception of multiple bodies of knowledge co-produced by researchers, practitioners and other stakeholders across various contexts (Newiga et al., 2019). In the ideal world, Mode 2 implies urban planning researchers and practitioners collaborating to co-produce research that is both scientifically rigorous and socially relevant to ensure robust policies and plans.

However, this is seldom the case in the complex and messy world of planning practice. Despite Mode 2, the urban planning literature points towards a gap between research and practice, with researchers and practitioners forming two distinct communities with different values and different understandings of and uses for research (Burton, 2018; Campbell, 2015; Forsyth, 2016; Goodman et al., 2022; Hurley et al., 2016; Porter, 2015). Researchers typically value rigour, whereas practitioners value relevance. How can urban planning research be both rigorous and relevant? This question is pertinent for Sustainable Development Goal (SDG)11, to make cities and human settlements inclusive, safe, resilient and sustainable, and SDG11.3 in particular, to enhance inclusive and sustainable urbanisation and capacity for participatory, integrated and sustainable human settlement planning and management (United Nations, 2023). As such, it is also indirectly pertinent to other SDGs within and beyond SDG11 that intersects urban planning, such as housing, transport, public spaces, climate action and water sensitivity.

Transdisciplinarity aims to combine scientific rigour and societal relevance. Transdisciplinarity can be defined as collaboration between academic role players, such as planning researchers or researchers from other disciplines, and non-academic role players, such as planning practitio-

---

J. du Toit (✉)
Department of Town and Regional Planning, University of Pretoria, Pretoria, South Africa
e-mail: jacques.dutoit@up.ac.za

A. Pieterse
Smart Places Cluster – South African Council for Scientific and Industrial Research, Pretoria, South Africa
e-mail: apieterse@csir.co.za

S. Mbatha
Department of Cooperative Governance and Traditional Affairs [COGTA], Pretoria, South Africa
e-mail: sandilemb@cogta.gov.za

© The Author(s) 2025
F. Frehse et al. (eds.), *Spatial Methods in Transdisciplinarity for Urban Sustainability*, Sustainable Development Goals Series, https://doi.org/10.1007/978-3-031-84367-9_1

ners and local communities, to research, learn about and work towards a common goal or structural change in a normative direction involving a complex societal or planning problem, such as urban sustainability (Cilliers et al., 2014, p. 261; Brink et al., 2018, p. 766). Transdisciplinarity therefore shifts from positivism and interpretivism to realism and pragmatism and assumes a more critical role for research methods, one in which methods are not only about rigour but also relevance to bring about systemic change towards urban sustainability.

Yet, how should urban planning researchers, practitioners and other stakeholders collaborate and conduct transdisciplinary research? In this chapter, 'spatial methods' are regarded as social research methods applied to questions involving spatial phenomena. This may include various methods that depict physical as well as social spaces, with the potential to transform such spaces towards greater sustainability. Instead of advocating specific methods, this chapter presents a holistic and flexible methodological framework for transdisciplinary urban planning. In this chapter, 'transdisciplinary urban planning' is regarded as academic and non-academic role players, from or beyond planning, collaborating to produce knowledge for urban planning purposes, especially in terms of addressing urban sustainability at local government level. This chapter first reviews the literature on transdisciplinarity for urban sustainability, after which the framework is presented and discussed. The heuristic framework serves to help urban planning stakeholders navigate transdisciplinarity and make more considered decisions for conducting transdisciplinary research, especially with regard to leveraging the transformative potential of research methods towards achieving urban sustainability. Practitioner reflections on the framework are provided using the example of Planning Support Science (PSSci) and a customised Planning Support System (PSS) for climate resilient planning at the local government level. The PSS was developed by the South African Council for Scientific and Industrial Research (CSIR) for eThekwini Municipality in Durban, South Africa.

1.2 A Review of Literature on Transdisciplinarity for Urban Sustainability

The urban planning literature cited above conceptualises the relationship between research and practice as a cultural gap between researchers and practitioners that limits the rigour and relevance of research. A recurring theme is that greater collaboration between researchers and practitioners may help bridge this gap, and that collaboration is *inter alia* a function of methodology. Transdisciplinarity is intended for this purpose; however, urban planning literature has largely overlooked this topic. We therefore reviewed literature across interdisciplinary fields such as urban and sustainability studies for examples of transdisciplinarity for urban sustainability, including successful or at least partially successful stakeholder collaborations. Instead of including or excluding particular examples, we reviewed articles published over the last decade with 'transdisciplinary' and 'urban sustainability' as our main keywords and grouped examples from these articles under four headings, including the Social Polis, African, cross-country and institutional examples. We subsequently describe these examples and synthesise their methodological implications into principles for a methodological framework.

1.2.1 Examples of Transdisciplinarity for Urban Sustainability

1.2.1.1 The Social Polis Example

'Social Polis' was an international transdisciplinary project between 2007 and 2010 that aimed to set a research agenda for urban social cohesion using an open social platform involving researchers, practitioners, policymakers and civil society, the majority of which were based in Europe. Setting the research agenda first involved a broader objective with high-priority themes, followed by a focused agenda that included two major societal challenges and five specific themes (Cassinari & Mouleart, 2015). The project was challenging for both practitioners and researchers. Practitioners found the themes abstract and removed from their normal experience of applied research to solve tangible problems, while researchers found it difficult to exchange the traditional linear approach of knowledge production for a cumulative-circular approach of mutual learning. Cassinari and Mouleart (2015) conclude that considering methodological aspects during the agenda setting can help address such challenges on a conceptual level.

A methodological framework that considers the contextual, epistemological and methodological dimensions of research can therefore help guide transdisciplinary research agendas. Social Polis is a predominantly European example, whereas examples elsewhere in the world, especially in Africa, face unique challenges.

1.2.1.2 African Examples

Transdisciplinarity for urban sustainability faces critical methodological challenges in an African context. Reflecting on the African Centre for Cities, Parnell and Pieterse (2016) put forward the notion of 'translational urban research praxis' to suggest a form of research beyond applied and even transdisciplinary research, which encompasses and integrates conception, design, execution, application and reflection. Together these form a 'singular research/practice process that is, by its nature, deeply political and locally embedded' (Parnell & Pieterse, 2016, p. 236). 'Translational' means directly engaging the transformation of urban practices through producing knowledge to inform change as well as knowledge about change itself, i.e. critical reflection. They, however, also point out a lack of methodological consensus, weak or missing data in the African context, and few researchers or practitioners having sufficient quantitative and qualitative skills. Yet they argue that integrating theory and method remains a meta-paradigmatic concern for urban sustainability. This holds three implications for a methodological framework. First, it suggests that a framework should include basic or fundamental research to critically reflect on change itself, in addition to applied or problem-driven research to inform change.

Second, it should be methodologically flexible to accommodate possible data and skills shortages, yet coherent to ensure some degree of integration between theory and method. Third, it should leverage the skills of researchers to improve the rigour of data collection and analysis.

Besides these implications, it should also be noted that transdisciplinary approaches from developed countries cannot necessarily be replicated to developing countries or settings characterised by informality, fluidity and social conflict. Van Breda and Swilling (2019) conducted a transdisciplinary case study between 2011 and 2016 in Enkanini, an informal settlement in Stellenbosch, South Africa. Enkanini formed when people occupied municipal land and erected informal dwellings. As it could take several years for the municipality to provide formal municipal services to the settlement, the question driving the project was: what could residents do while awaiting services? Over the course of 3 years, the research team in collaboration with a loose network of residents implemented three small-scale experiments in alternative service delivery, including electricity, sanitation and waste management. Reflecting on their experiences, the authors argue for 'emergent transdisciplinary design research', i.e. the research is designed as it unfolds, transforms and emerges *from* and *within* the local context. Yet they also suggest that contextually relevant guiding logics and principles are still needed for conducting transdisciplinary research. A number of principles are formulated, two of which have implications for a methodological framework, including 'allowing for emergence' and 'absorbing complexity'. Research planning and design, although a collaborative effort, should, if possible, preferably be led by local practitioners or community members who are better positioned to allow and absorb such emergence and complexity, i.e. anticipating and working with uncertainty and unexpected circumstances. A methodological framework should therefore also be flexible with regard to research designs and methods, and incorporate some of the broader principles of transdisciplinary research. Van Breda and Swilling (2019, pp. 827–828), however, make an important conceptual distinction between methodology, i.e. the critical reflection on the reasoning or logic behind a chosen method, and the method itself, whereas method is instrumental and does not reveal why or how it should be used.

Although this chapter argues for methodological flexibility and plurality, the complexity and fluidity of transdisciplinary and sustainability problems still require careful consideration of the logic behind a chosen method. A methodological framework should therefore provide guidance on how a prototypical research design or method is more-or-less related to the research context, purpose and logic so that stakeholders can be more considered in their methodological decisions.

1.2.1.3 Cross-Country Examples

Transdisciplinary research is further complicated when conducted comparatively between countries and cities. Mistra Urban Futures (MUF) is an international research centre that facilitates applied transdisciplinary research towards implementing the SDGs at city level through a series of 'Local Interaction Platforms' in Gothenburg, Malmö, Lund and Stockholm (Sweden), Sheffield and Manchester (UK), Cape Town (South Africa), Kisumu (Kenya), Buenos Aires (Argentina) and Shimla (India) (Simon et al., 2018; Valencia et al., 2019). Some of the lessons and generalisable principles for practitioners and researchers include being (1) locally appropriate and embedded, (2) open to change and renewal and (3) able to straddle disciplines, bridge the gap between research and practice and accept that there is no single right way of doing transdisciplinary research. This highlights the potential of planning researchers to help bridge this gap, considering that planning is foremost a practice-oriented discipline. It also suggests the need for a flexible methodological framework that, rather than advocating specific methods, allows for various methods that may be applicable. Another important lesson is that the delineation of urban boundaries for transdisciplinary research has implications in terms of which stakeholders and urban processes are being included or excluded. Again, local practitioners who are better positioned to grasp and appreciate these dynamics should perhaps lead the research planning and design, which includes delineating the research. As part of accountability towards stakeholders, Valencia et al. (2019) suggest effective SDG coordination mechanisms, which include monitoring and evaluation using both qualitative and quantitative assessments of progress with measurable indicators that are contextually relevant. A methodological framework should therefore include evaluation research, considering also that it is a well-established form of applied research in urban planning. The role of researchers in building capacity for rigorous data collection and analysis is critical. Simon et al. (2018) also draw attention to the role of researchers in terms of public engagement and science communication, i.e. to expose practitioners and stakeholders to new perspectives through succinct literature reviews, thereby improving the relationship between research and practice to achieve greater social impact, which also relates to the European Union's efforts towards Responsible Research and Innovation.

In essence, MUF also provides a critical reflection on transdisciplinary research. This again suggests the need to include basic or fundamental research in a methodological framework to create reflexive knowledge.

Comparative transdisciplinary research across African countries also confronts urban sustainability challenges. Thondhlana et al. (2021) reflect on two transdisciplinary projects related to climate change and household energy use that were implemented in Kumasi (Ghana), Durban and Makhanda (South Africa) and Harare (Zimbabwe), as part of the Leading Integrated Research for Agenda 2030 in Africa. Two lessons were that asymmetrical power dynamics between stakeholders may influence outcomes, and that partnerships therefore require clearly defined agreements to govern structure and process.

This is perhaps contrary to notions of 'emergent' and 'adaptable' approaches in the examples above. Yet it suggests that establishing a clearer role division between planning researchers and practitioners, based on the methodological strengths each party brings to a project, may help mitigate possible asymmetrical power dynamics or the need for rigid agreements.

1.2.1.4 Institutional Examples

Transdisciplinary research can be challenged by institutional aspects influencing teleology and methodology. The Institute for Sustainable Urban Development (ISU) was a collaborative project between Malmö University and the municipality to facilitate research and planning collaboration for urban sustainability and education. The ISU resulted in different internal and external understandings of the institution as either a dependent bilateral but invisible project, or as a high-profile independent institute. Organisers, however, felt it did not result in a 'self-organising network across sectors', and that it was teleologically aimless (Mehling & Kolleck, 2019). Patel (2022) similarly reflects on a city-university knowledge partnership for urban sustainability in Cape Town, arguing that although 'third spaces' between academia and practice are generally accepted as integral to transdisciplinarity, researchers occupying such spaces may be peripheralised from their academic institutions, limiting the potential of transdisciplinary research to affect urban change. The notion of a 'portal' is put forward to suggest the need for dynamic exchange between researchers and their institutions to affect urban change. Brink et al. (2018) provide a self-assessment of a joint project across four universities and seven Swedish municipalities, titled Ecosystems Services Concept at the Municipal Level, and identify lessons for supporting university-municipality collaboration. One of these is that greater attention to the purpose of stakeholder participation, whether functional, deliberative or emancipatory, can help justify and clarify roles. They also suggest that researchers clearly articulate their epistemological, ontological and methodological positions.

This, again, suggests the need for a holistic framework that provides methodological coherence between the different dimensions of research, and that clarifies the role of researchers and practitioners in terms of participation and methodological contribution.

1.2.2 Principles for a Methodological Framework

The methodological implications from the examples above can be synthesised into four principles for a methodological framework. Ideally, there should be:

- Methodological coherence between the different dimensions of research
- Methodological role division between stakeholders
- Contextuality, flexibility and plurality with regard to research designs and methods
- Critical reflection on transdisciplinary urban planning

The literature suggests that for transdisciplinary research to be scientifically rigorous, a degree of methodological coherence is necessary, which requires a matrix-like framework that aligns concomitant research contexts, paradigms, purposes, designs and methods. Greater clarity regarding the role of different stakeholders is called for, which requires a framework that leverages the respective positions and methodological strongpoints of different stakeholders. The complexity of urban sustainability challenges requires a framework that allows stakeholders to choose and experiment with research designs and methods that are contextual, flexible and plural. The need to not only affect change but also to learn about change requires a framework that allows critical reflection on transdisciplinary urban planning. Each of these four principles is considered in the subsequent framework.

1.3 A Methodological Framework for Transdisciplinary Urban Planning

Methodological frameworks are useful to help stakeholders navigate complexity, especially transdisciplinarity for urban sustainability. They furthermore help to position research methods and how these may apply to spatial phenomena as 'spatial methods'. While methodological frameworks (Wiek & Lang, 2016) and principles (Kudo et al., 2019; Von Wehrden et al., 2017) have been proposed for sustainability research in general, the literature lacks specific frameworks for transdisciplinary urban planning. The framework proposed here incorporates the four principles from the review above and draws on existing typologies of designs for planning research (Du Toit, 2015). The framework (see also Table 1.1) is structured around four dimensions of social science research, including:

- The sociological dimension, i.e. the research context and audience
- The epistemological and ontological dimension, i.e. the research paradigm and focus
- The teleological dimension, i.e. the research purposes, principles or logics
- The methodological dimension, i.e. the research planning, design and methods

Table 1.1 shows the framework.

The research context pertains to whether research is primarily applied or basic. Applied research involves problem-driven or solution-oriented research primarily for a practitioner or community audience, whereas basic research involves fundamental research primarily for an academic audience to improve the intrinsic understanding about transdisciplinary urban planning (Neuman, 2020). Planners are familiar with applied research, while transdisciplinarity for urban sustainability is typically applied at local government level (see, e.g. Cilliers et al., 2014; Brink et al., 2018; Thondlana et al., 2021). A distinction, however, is made between applied research primarily involving practitioners, which tends towards intervention and evaluation research (the two main forms of applied planning research) and applied research primarily involving communities, which tends towards participatory action research (PAR), and community-based research (Leedy & Ormrod, 2021; Neuman, 2020). Westin and Joosse (2022, pp. 392 & 401) point out that planners are well positioned to mediate between experts and citizen authority, and to facilitate the process of participation. Several authors also point out the need for researchers to critically reflect on transdisciplinary projects and to conduct basic research *on* planning to theorise experiences and change, which can be either in tandem to or after a transdisciplinary project (Parnell & Pieterse, 2016; Brink et al., 2018, pp. 767–769; Simon et al., 2018; Lawrence, 2021, p. 205). Applied and basic research therefore form a continuum to re-enforce knowledge *for* urban sustainability and knowledge *on* transdisciplinary urban planning itself.

Considering its applied focus, intervention and evaluation research tend to be non-paradigmatic or pragmatic, as the purposes are to intervene in urban settings and to evaluate such interventions to create instrumental knowledge for planning. As in the social sciences, PAR in the context of transdisciplinary urban planning is associated with realism (Brink et al., 2018, p. 779), also known as critical social science (Neuman, 2020) or pragmatism. According to Popa et al. (2015, pp. 47–48), pragmatism 'distances itself from value neutrality and value relativism by conceiving knowledge production as a social and reflexive process, whereby criteria of scientific credibility and legitimacy are jointly defined within a community of inquiry'. The focus is on research *for* the community to create either instrumental or reflexive knowledge. Realism, pragmatism and involvement of the community imply democratic principles such as participation, deliberation and eventual transformation or emancipation. Transdisciplinarity implies praxis, i.e. creating knowledge through action and experimentation to see if theory works in practice and to refine theory if necessary. The actual benefits for the community, however, need to be continuously discussed (Brink et al., 2018, p. 779). Although basic research may involve empirical case studies to critically reflect on transdisciplinary urban planning, research that focuses *on* planning to create reflexive knowledge is more likely meta-paradigmatic, especially if the emphasis is on non-empirical work such as theory construction (Van Breda & Swilling, 2019, p. 828), methodological reflection (Von Wehrden et al., 2017, p. 35; Van Breda

Table 1.1 A methodological framework for transdisciplinary urban planning

Sociological dimension (research context)	Epistemological and ontological dimension (research paradigm and focus)	Teleological dimension (research purposes, principles or logics)	Methodological dimension (Research planning and design)	(Research methods)
Applied (problem-driven/solution-oriented) research primarily involving *practitioners* as key stakeholders, particularly in the context of practice and local government	Non-paradigmatic or pragmatic Research *for* planning to create instrumental knowledge	Intervention Evaluation	LED BY PRACTITIONERS *Intervention research* Building/site/settlement analysis (incl. Environmental scanning) Plan/policy analysis Normative/exploratory forecasting *Evaluation research* Clarificatory evaluation Implementation evaluation / Programme monitoring Impact evaluation	LED BY RESEARCHERS Literature reviews, science communication (engagement) and tools such as Planning Support Systems, to offer practitioners new perspectives and evidence Various quantitative and/or qualitative data collection, analysis and mapping methods that are (1) pragmatic, (2) experimental, (3) longitudinal, (4) ethical and (5) driven by practitioners' policy and planning imperatives
Applied (problem-driven/solution-oriented) research primarily involving *communities* as key stakeholders in a localised context, with planners as mediators and facilitators	Realist or pragmatic Research *for* communities to create instrumental/reflexive knowledge	Participation/ deliberation Action/ experimentation (praxis) Transformation/ emancipation	LED BY COMMUNITIES *Participatory (action) research* (e.g. functional, deliberative or transformative/ emancipatory) *Community-based research* Course-based action research Community-based participatory research	LED BY RESEARCHERS Literature reviews and science communication (engagement) to offer communities new perspectives and evidence Various quantitative and/or qualitative data collection and analysis methods that are (1) emergent, (2) iterative, (3) collaborative and (4) ethical
Basic (fundamental) research conducted by *researchers* in an academic context	Meta-paradigmatic Research *on* planning to create reflexive knowledge	Critical reflection	EMPIRICAL *Case studies* of transdisciplinary urban planning NON-EMPIRICAL *Theory construction* *Methodological studies* *Normative argumentation*	Various empirical case study methods

& Swilling, 2019, p. 826) or normative argumentation.

The methodological dimension is divided into research planning, design and methods. 'Research design' refers to the prototypical study and can be defined as a logical plan involving strategic decisions to maximise research purposes, applications or the validity of findings (Du Toit, 2015, p. 61). Prototypical evaluation designs, for example, include clarificatory, implementation and impact evaluations. Prototypical PAR designs include participatory or action designs and community-based research (Brink et al., 2018, p. 779; Leedy & Ormrod, 2021). 'Research methods' refers to data collection and analysis procedures as part of research design. Several authors, however, point out issues around stakeholder roles in transdisciplinary research, and that this is often a matter of differential expertise and a need for greater clarity in terms of process and methodological contribution (Mitchell et al., 2015; Popa et al., 2015; Parnell & Pieterse, 2016, pp. 241–242; Brink et al., 2018, pp. 774–775 & 781; Simon et al., 2018, pp. 484–487; Mehling & Kolleck, 2019; Valencia et al., 2019, pp. 6–7). The framework therefore proposes a collaborative partnership between planning researchers and practitioners with a role division in terms of their respective methodological contribution, with practitioners ideally leading the research planning and design (establishing relevance and aligning research activities with local policy and planning imperatives), and researchers leading the methods (establishing rigour by ensuring reasonably valid and reliable (quantitative) or credible and trustworthy (qualitative) data collection and analysis procedures) (Von Wehrden et al., 2017, pp. 39–40). The intention is not to create an asymmetrical relationship in terms of methodology, but rather to leverage the relative positions and strengths that each party brings to the collaboration. The intention is also not that research planning and design should be left completely for practitioners or communities. More structured designs such as surveys or field experiments, in which the validity of findings tend to be strongly tied to the planning and design thereof, may still benefit from researchers' expertise and inputs (as the CSIR and eThekwini authors similarly remark in Sect. 1.4.2).

Researchers may provide succinct literature reviews in response to specific practitioner or community needs, thereby enabling otherwise arcane research to have greater societal impact through what Simon et al. (2018, pp. 487 & 494–495) refer to as 'engagement' or 'science communication'. Researchers may also develop and facilitate tools such as PSS to assist practitioners with planning and decision-making. Considering the complex and fluid nature of transdisciplinary research, the framework proposes various rather than specific methods, as long as methods are appropriate and aligned with other research considerations considering criteria for good qualitative or quantitative research (Von Wehrden et al., 2017, pp. 37–39). For applied research involving practitioners, it is important that methods are pragmatic, ethical and driven by practitioners' policy and planning imperatives (Parnell & Pieterse, 2016, p. 240). Where applicable, methods should also be experimental to determine change (Wiek & Lang, 2016, p. 32; Von Wehrden et al., 2017, p. 39) or longitudinal to track change (Von Wehrden et al., 2017, p. 40). For applied research involving communities, it is important that methods are emergent, iterative, collaborative and ethical.

The heuristic framework serves to help stakeholders navigate transdisciplinarity and make more considered decisions when conceptualising, planning and designing research for transdisciplinary urban planning. The framework is not meant to be prescriptive or rigid. Additional methodological considerations beyond those outlined in the framework may be taken into account. Furthermore, elements from one section may be applicable to other sections. For example, participation as a purpose or principle, although typically associated with action research involving communities, may also apply to intervention research involving practitioners. Still, the framework provides some methodological coherence, i.e. how key methodological considerations more-or-less fit together, which is important for research to be both rigorous and relevant (Von Wehrden et al., 2017, p. 37). Practitioners subsequently reflect on the framework to illustrate how the framework may apply to a tangible example of transdisciplinary urban planning.

1.4 Practitioner Reflections on the Framework Using the Example of Planning Support Science (PSSci) and a Customised Planning Support System (PSS)

1.4.1 The Concept of PSSci and PSS

PSSci is the application of scientific knowledge, methods and tools to support decision-making in urban planning. PSSci recognises the multi- and trans-disciplinary nature of urban planning and aims to integrate various scientific disciplines to support practitioners. It emphasises the collection and analysis of data, modelling, simulation, participatory decision making and evaluation practices. PSSci is concerned with the overall theoretical and methodological framework that guides the use of scientific knowledge in planning processes. Recent developments in PSSci highlight the need for closer collaboration between government, knowledge institutions, industry and civil society to better integrate the governance, application and instrumentation components of PSSci (Geertman & Stillwell, 2020).

PSSs, on the other hand, refer to the technological tools and computer-based systems that aid decision-making in planning processes. PSSs are interactive software systems designed to assist planners in analysing data, visualising information, modelling scenarios and evaluating planning alternatives. These systems often incorporate geographic information systems (GIS), data management tools, modelling software and visualisation platforms (Geertman, 2015). PSSci concepts and methods are therefore applied in PSS. The current focus, however, has shifted from PSS as mere instruments and technologies (the means of support) to the position of PSS within the context of planning (the goals of support).

> The focus has changed to one in which we question how PSS can be embedded in a specific application field, what role can PSS play given the particular governance procedures in force, what methodological relationships exist between PSS and other associated instruments, and what impact does the contextual setting have on PSS design and use. (Geertman & Stillwell, 2020, p. 1329)

Planning issues are often complex and multidimensional, requiring integrated responses. PSSs provide knowledge for integrated responses and support the communication and analysis of that knowledge (Vonk & Geertman, 2008). PSSs typically offer an integrated framework of planning theory, methods, instruments, data, information and knowledge (Geertman, 2015). Vonk and Geertman (2008, p. 155) argue that PSS can support planners in 'developing well founded plans for an increasingly complex socio-physical environment in increasingly complex planning process environments'. PSS and the processes in which they are developed serve as a transdisciplinary bridge between planning research and practice. Zheng and Sieber (2020) argue that it is the inter- and trans-disciplinary collaborations that truly drive urban innovation. Reflecting on the methodological framework in Table 1.1, PSSci serves as an example of applied research involving practitioners. The concern in PSSci with an overall theoretical and methodological framework to guide the use of scientific information in planning, reflects the methodological dimension in Table 1.1, particularly the research planning and design. PSS, on the other hand, reflects research methods, considering the role of researchers in developing tools such as PSS. PSSci and PSS together serve as a methodological vehicle for intervention and evaluation research.

1.4.2 eThekwini GreenBook MetroView

In 2021, a collaboration between the CSIR and eThekwini Municipality led to the development of a customised PSS for climate-responsive strategic planning, known as the eThekwini GreenBook MetroView. The MetroView forms part of the larger GreenBook initiative that provides evidence for local government in South Africa through a functional web-based platform to inform the planning and development of climate-resilient settlements. As part of the GreenBook, the MetroView includes two tools, the Climate Risk Profile Tool and the Climate Actions Tool. The Climate Risk Profile Tool, depicted in Fig. 1.1, provides a baseline and future profile of climate and climate change, the likelihood of certain climate hazards to occur, the exposure and vulnerability of people and infrastructure, and combined risk. The profile information is quantified and spatialised at three planning scales including municipal, regional and ward levels. The Climate Actions Tool provides customised adaptation and mitigation actions to be mainstreamed into municipal plans and strategies to reduce exposure and vulnerabilities and build long-term resilience. The 2022 Durban Climate Change Strategy (eThekwini Municipality, 2022a) informs the inclusion and structure of the different climate change adaptation and mitigation actions. These tools therefore provide data-driven climate risk assessments of eThekwini using a combination of GIS, modelling and visualisation techniques. Together, the two tools offer a strong evidence base in the form of communication and analysis to assist practitioners, policymakers and decision-makers to effectively inform and add value to climate change adaptation and strategic planning. The tools also provide a foundation for developing a shared understanding of the potential impacts of climate change and the decisions needed to respond to them (CSIR, 2023a).

Reflecting on the methodological framework in Table 1.1, the development of the eThekwini GreenBook MetroView was problem driven, solution oriented and customised for eThekwini to support evidence-based strategic planning for a climate-resilient city. Therefore, the focus of the research was to create both instrumental and reflexive knowledge for urban sustainability. Instrumental knowledge was created through the development of an evidence base *for* planning and decision-making in the city, whereas reflexive knowledge was created *on* planning around broader systemic dynamics, power structures and ethical considerations that shape decision-making processes. Both types of knowledge benefit stakeholders in the research process. Instrumental knowledge provides practical guidance and decision support to eThekwini stakeholders, while it supports technical expertise and the optimisation of research for the CSIR researchers. Reflexive knowledge offers eThekwini stakeholders a critical lens through which they can better understand and shape the implications of their decisions, while for CSIR

Fig. 1.1 An excerpt of the Climate Risk Profile from the eThekwini GreenBook MetroView. (Source: CSIR, 2023b)

researchers it offers a normative and evaluative lens to consider the implications of the research process and outputs.

The CSIR, the research partner and developer of the GreenBook, mostly conducts applied research in the local government sector. Since the development of this PSS in 2016 (Van Niekerk et al., 2020), the CSIR has designed and updated its methodologies for the development of the PSS to be in line with the latest research in urban planning, adaptation and climate change sciences. A reliable theoretical and methodological base for the development and refinement of the PSS was an essential part of the process. The methodology for the eThekwini GreenBook MetroView is in line with the latest Intergovernmental Panel on Climate Change Sixth Assessment Report (IPCC, 2021). Through close collaboration with eThekwini, the methodology was adapted to fit the data available to the research team.

The successful development of the eThekwini GreenBook MetroView was largely dependent on stakeholder buy in and collaboration at municipal level. It leveraged existing initiatives of city-level data integration, which demonstrated institutional capacity to use data for insight-driven decision-making. The development process was also enabled by other mechanisms that made resources available to support this transdisciplinary collaboration, which may otherwise have been stifled by procurement systems and other local government legislative requirements. Development of this PSS was facilitated by funds made available from the CSIR as a research partner, the Absa Group as a private sector stakeholder and the National Treasury's City Support Programme as a government stakeholder.

To ensure that the eThekwini GreenBook MetroView was internalised within the local government machinery, a series of internal workshops were conducted. The purpose of these workshops was threefold; first, to identify and develop the most appropriate methods and data sources for the various components of the Climate Risk Profile Tool; second, to validate findings; and third, to ensure buy in and ownership within the municipality. The workshops involved a multidisciplinary team of CSIR researchers, as well as municipal officials and practitioners from various departments. A transdisciplinary co-development approach was followed to ensure transparency in the development of the PSS and its tools and to avoid municipal silos or under-utilisation. The PSS was eventually located within existing municipal platforms to further strengthen its ownership. The PSS is currently shared through the municipality's eThekwini Strategic Hub data platform (eThekwini Municipality, 2022b). The most significant co-development between practitioners and researchers is related to establishing and developing the research design and methods. Thus, although the methodological framework in Table 1.1 suggests that practitioners and researchers should lead the research design and methods respectively, the eThekwini experience suggests that co-development is simultaneously important (as the academic author similarly remarks in Sect. 1.3). It may also at times be necessary to go beyond intentional designs or purposes such as intervention, evaluation and participation and allow for flexibility and unintended or serendipitous outcomes of the research. The project experienced significant delays as the process of co-development was more time intensive than the team originally anticipated, which necessitated effective and transparent communication between researchers, practitioners and funders. Although not an aspect in the framework, users of the framework should emphasise open and regular communication to ensure that all stakeholders are aware of potential delays, changes or serendipitous outcomes.

The eThekwini GreenBook MetroView has also been workshopped with civil society and community stakeholders to support grassroot planning processes. Existing relationships between the municipality and these stakeholders determine whether instruments such as PSS are accepted or rejected. In the context of a functional relationship, stakeholders are more likely to accept the use of such instruments for their own planning purposes. In some instances, they may even provide critical data required to enhance instruments. A trust deficit may, however, lead to a rejection of an instrument. Thus, the methodological framework in Table 1.1 highlights the importance of community stakeholders leading the planning and design of participatory and community-based research, which would include using a PSS such as the eThekwini GreenBook MetroView for grassroot planning processes.

1.5 Conclusion

Considering the need for greater collaboration between researchers and practitioners to achieve urban sustainability, this chapter presents a methodological framework for transdisciplinary urban planning. The literature as yet offers little with regard to such a framework. Literature from interdisciplinary fields, such as urban and sustainability studies, offers examples of transdisciplinarity for urban sustainability, from which we synthesised principles for a methodological framework. The framework is structured around four dimensions of social science research (i.e. sociological, epistemological, teleological and methodological) to help align context, paradigm, purpose and research planning, design and methods. Applied research involves practitioners and communities as primary stakeholders to create instrumental knowledge *for* planning and communities. Prototypical designs particularly suited to transdisciplinary urban planning include intervention, evaluation and participatory research, in which practitioners and communities, considering their unique position, should ideally lead the research planning and design by coordinating research efforts and practical gains. Researchers, on the other hand, should offer succinct literature reviews and ideally lead methods for data collection and analysis considering their expertise. Rather than advocating specific methods, the framework encourages a plural and flexible approach, as long as these methods are appropriate and responsive to the needs of practitioners and communities. It is about the pragmatic and ethical application of a method rather than the method itself. Basic research is conducted by researchers parallel or subsequent to applied research to create reflexive knowledge *on* planning. This is typically done through meta-paradigmatic case studies of transdisciplinary urban planning to contribute towards theory, methods or norms for future transdisciplinary research.

Reflections on the framework are based on the example of PSSci and a customised PSS. PSS facilitates transdisciplinary urban planning by aligning practitioner needs with research methods and techniques. This marks a departure from highly technical tools such as GIS that do not always meet the planning needs at local government level. The CSIR and eThekwini Municipality collaborated on the eThekwini GreenBook MetroView, a web-based instrument for strategic planning in terms of climate responsiveness at local government level. From a practitioner's perspective involved with this PSS, the framework may help to conceptualise and design transdisciplinary urban planning research while leveraging unique practitioner and researcher capabilities. The framework, however, should not be prescriptive and cannot foresee all the contextual and unique challenges that a transdisciplinary project necessarily begets. In South Africa, transdisciplinarity at local government level is hindered by complex political and governance dynamics. Local government is constrained by municipal service backlogs and the lack of resources to respond. Legislation related to procurement processes also challenges researcher-practitioner collaboration. These challenges stifle practitioners' ability to collaborate and innovate, while transdisciplinarity and urban sustainability simply become secondary under such circumstances. A methodological framework can therefore at best propose plurality and flexibility with regard to methods to help mitigate such challenges. Even then, transdisciplinarity at local government level is largely dependent on institutional maturity and the willingness and capability to break internal silos and incorporate research into planning and decision-making for urban sustainability.

These reflections on the framework raise questions beyond methodology around the societal impact of transdisciplinary urban planning, especially in terms of urban sustainability at local government level. Transdisciplinarity implies praxis, but considering the challenges, especially at local government level, what is the actual impact, and how should it be evaluated? While the framework suggests that researchers should critically reflect on cases of transdisciplinarity, this is more likely to focus on planning processes, and not necessarily the impact thereof. There may also be a difference between evaluating the impact of a planning intervention and evaluating the impact of the research behind the intervention. As evaluating the impact of transdisciplinary research is probably beyond the framework in its current form, researchers may have to draw on science and technology studies, particularly the field of research evaluation. Notions such as productive interactions and impact pathways (Muhonen et al., 2020) may be useful to evaluate the impact of transdisciplinary research, including the role of methods in bringing about change. Hansson and Polk (2018) argue that impact depends on the quality of the research process, particularly in terms of how practitioner motivation, participant openness or flexibility and in-depth exchanges of expertise contribute to research that is internally relevant, credible and legitimate. These attributes, however, should also be related to external dynamics, institutional factors and political context. Urban planning has yet to see systematic evaluations of transdisciplinary research, which may in turn serve to refine methodological frameworks for transdisciplinary urban planning.

As a closing thought, we consider how our positionality may have played a role in constructing the framework and reflecting on it. The academic author, who trained as a planner and subsequently a methodologist, may have a more purist take on methodology and prefers the application of prototypical social research designs and methods in urban planning research, albeit in a pragmatic manner. The CSIR and eThekwini authors, who actively participated in the development of the GreenBook MetroView, may be predisposed towards emphasising the successes and positive outcomes of the process. This predisposition is driven by a personal investment in the success of the project. As a researcher and practitioner respectively, they recognise their need to balance the interests of various stakeholders, including the funders, the municipality and other stakeholders, which may influence the framing of their narrative.

References

Brink, E., et al. (2018). On the road to 'research municipalities': Analysing transdisciplinarity in municipal ecosystem services and adaptation planning. *Sustainability Science, 13*, 765–784.

Burton, P. (2018). Striving for impact beyond the academy? Planning research in Australia. In T. W. Sanchez (Ed.), *Planning knowledge and research* (pp. 51–65). Routledge.

Campbell, H. (2015). It takes more than just looking to make a difference: The challenge for planning research. In E. A. Silva, P. Healey, H. Neil, & P. Van den Broeck (Eds.), *The Routledge handbook of planning research methods*. Routledge.

Cassinari, D., & Moulaert, F. (2015). Enabling transdisciplinary research on social cohesion in the city. In E. A. Silva, P. Healey, H. Neil, & P. Van den Broeck (Eds.), *The Routledge handbook of planning research methods* (pp. 414–425). Routledge.

Cilliers, S., et al. (2014). Sustainable urban landscapes: South African perspectives on transdisciplinary approaches. *Landscape and Urban Planning, 125*, 260–270.

CSIR. (2023a). *eThekwini municipality climate risk profile*. Available at: https://ethekwini-riskprofile.greenbook.co.za/?view=about-the-tool. Accessed 30 May 2023

CSIR. (2023b). *eThekwini Municipality Climate Risk Profile*. [Online] Available at: https://ethekwini-riskprofile.greenbook.co.za/hazards?view=flooding&subview=flood+likelihood. Accessed 14 June 2023

Du Toit, J. (2015). Research design. In E. A. Silva, P. Healey, H. Neil, & P. Van den Broeck (Eds.), *The Routledge handbook of planning research methods* (pp. 61–73). Routledge.

eThekwini Municipality. (2022a). *Durban Climate Change Strategy*. Available at: https://www.durban.gov.za/storage/Documents/Climate/DCCS_Strategy.pdf. Accessed 13 June 2023

eThekwini Municipality. (2022b). *eThekwini Strategic Hub: Our single source of truth*. Available at: https://strathub.durban.gov.za/. Accessed 30 May 2023

Forsyth, A. (2016). Investigating research. *Planning Theory & Practice, 17*(3), 467–471.

Geertman, S. (2015). Planning support systems (PSS) as research instruments. In E. A. Silva, P. Healey, N. Harris, & P. Van den Broeck (Eds.), *The Routledge handbook of planning research methods* (pp. 322–334). Routledge.

Geertman, S., & Stillwell, J. (2020). Planning support science: Developments and challenges. *Environment and Planning B: Urban Analytics and City Science, 47*(8), 1326–1342.

Goodman, R., Freestone, R., & Burton, P. (2022). Planning practice and academic research: Views from the parallel worlds. *Planning Practice & Research, 37*(4), 497–508.

Hansson, S., & Polk, M. (2018). Assessing the impact of transdisciplinary research: The usefulness of relevance, credibility, and legitimacy for understanding the link between process and impact. *Research Evaluation, 27*(2), 132–144.

Hurley, J., Lamker, C. W., & Taylor, E. J. (2016). Exchange between researchers and practitioners in urban planning: Achievable objective or a bridge too far? *Planning Theory & Practice, 17*(3), 447–473.

IPCC. (2021). *Climate change 2021: The physical science basis. Contribution of working group I to the sixth assessment report of the intergovernmental panel on climate change*. Cambridge University Press.

Kudo, S., et al. (2019). Moving towards transdisciplinarity: Framing sustainability challenges in Africa. In M. Nagao, J. L. Broadhurst, S. Edusah, & K. G. Awere (Eds.), *Sustainable development in Africa: Concepts and methodological approaches* (pp. 1–14). Denver: Spears Media Press.

Leedy, P. D., & Ormrod, J. E. (2021). *Practical research: Planning and design* (12th ed.). Pearson.

Lawrence, R. J. (2021). *Creating Built Environments: Bridging Knowledge and Practice Divides*. Routledge.

Mehling, S., & Kolleck, N. (2019). Cross-sector collaboration in higher education institutions (HEIs): A critical analysis of an urban sustainability development program. *Sustainability, 11*(4982), 1–24.

Mitchell, C., Cordell, D., & Fam, D. (2015). Beginning at the end: The outcome spaces framework to guide purposive transdisciplinary research. *Futures, 2015*, 86–96.

Muhonen, R., Benneworth, P., & Olmos-Peñuela, J. (2020). From productive interactions to impact pathways: Understanding the key dimensions in developing SSH research societal impact. *Research Evaluation, 29*(1), 34–47.

Neuman, W. L. (2020). *Social research methods: Qualitative and quantitative approaches* (8th ed.). Pearson.

Newiga, J., et al. (2019). Linking modes of research to their scientific and societal outcomes. Evidence from 81 sustainability-oriented research projects. *Environmental Science and Policy, 101*, 147–155.

Parnell, S., & Pieterse, E. (2016). Translational global praxis: Rethinking methods and modes of African urban research. *International Journal of Urban and Regional Research, 40*(1), 236–246.

Patel, Z. (2022). The potential and pitfalls of co-producing urban knowledge: Rethinking spaces of engagement. *Methodological Innovations, 15*(3), 374–386.

Popa, F., Guillermin, M., & Dedeurwaerdere, T. (2015). A pragmatist approach to transdisciplinarity in sustainability research: From complex systems theory to reflexive science. *Futures, 65*, 45–56.

Porter, L. (2015). Unsettling comforting deceits: Planning scholarship, planning practice and the politics of research impact. *Planning Theory & Practice, 16*(3), 293–296.

Simon, D., et al. (2018). The challenges of transdisciplinary knowledge co-production: From unilocal to comparative research. *Environment & Urbanization, 30*(2), 481–500.

Thondhlana, G., et al. (2021). Facilitating urban sustainability through transdisciplinary (TD) research: Lessons from Ghana, South Africa, and Zimbabwe. *Sustainability, 13*(6205), 1–18.

United Nations. (2023). *THE 17 GOALS | sustainable development*. Available at: https://sdgs.un.org/goals. Accessed 14 June 2023

Valencia, S. C., et al. (2019). Adapting the sustainable development goals and the new urban agenda to the city level: Initial reflections from a comparative research project. *International Journal of Urban Sustainable Development, 11*(1), 4–23.

Van Breda, J., & Swilling, M. (2019). The guiding logics and principles for designing emergent transdisciplinary research processes: Learning experiences and reflections from a transdisciplinary urban case study in Enkanini informal settlement, South Africa. *Sustainability Science, 14*, 823–841.

Van Niekerk, W., Pieterse, A., & Le Roux, A. (2020). Introducing the green book: A practical planning tool for adapting south African settlements to climate change. *Town and Regional Planning, 77*, 103–119.

Von Wehrden, H., Luederitz, C., Leventon, J., & Russell, S. (2017). Methodological challenges in sustainability science: A call for method plurality, procedural rigor and longitudinal research. *Challenges in Sustainability, 5*(1), 35–42.

Vonk, G., & Geertman, S. (2008). Improving the adoption and use of planning support systems in practice. *Applied Spatial Analysis and Policy, 1*, 153–173.

Westin, M., & Joosse, S. (2022). Whose knowledge counts in the planning of urban sustainability? *Planning Theory & Practice, 23*(3), 388–405.

Wiek, A., & Lang, D. J. (2016). Chapter 3: Transformational sustainability research methodology. In H. Heinrichs, P. Martens, G. Michelsen, & A. Wiek (Eds.), *Sustainability science* (pp. 31–41). Springer Science+Business Media.

Zheng, Z., & Sieber, R. (2020). Planning support systems and science beyond the smart city. In S. Geertman & J. Stillwell (Eds.), *Handbook of planning support science* (pp. 199–212). Edward Elgar.

Jacques du Toit holds a PhD in methodology and is an Associate Professor in the Department of Town and Regional Planning, University of Pretoria, South Africa. His substantive research interest is in urban sustainability, particularly water-sensitive planning and design, and his procedural research interest is in evidence-based urban planning. He teaches infrastructure planning, policy analysis and research methods, and coordinates final year research report and mini-dissertation projects. Part time he also conducts applied research and serves on various institutional and faculty level committees.

Amy Pieterse holds an MA in Town and Regional Planning from the University of Pretoria and is a Senior Knowledge Applicator in the Smart Places Cluster at the Council for Scientific and Industrial Research, South Africa. Her expertise includes climate change adaptation planning, mainstreaming, policy analysis, and local government planning. She has a specific research interest in how local governments use research and evidence in planning and policy, with a focus on adaptation. Amy is an innovator, developer and manager of the GreenBook (greenbook.co.za), a practical online planning support system for the development of climate-resilient South African cities and towns.

Sandile Mbatha is the National Chief Data Officer at the Department of Cooperative Governance and Traditional Affairs, South Africa, where he is responsible for establishing the National Strategic Hub. The core of his work is to integrate, transform, analyse and visualise country-level data to extract value for informing strategic decision-making on planning, resource allocation, implementation and monitoring. He is responsible for building the National Strategic Hub which is the country's single source of truth. Previously he served as Senior Manager, responsible for Data, Research and Policy Advocacy in eThekwini Municipality where he supported evidence-based policy development processes through city-wide research and strategic planning.

Open Access This chapter is licensed under the terms of the Creative Commons Attribution 4.0 International License (http://creativecommons.org/licenses/by/4.0/), which permits use, sharing, adaptation, distribution and reproduction in any medium or format, as long as you give appropriate credit to the original author(s) and the source, provide a link to the Creative Commons license and indicate if changes were made.

The images or other third party material in this chapter are included in the chapter's Creative Commons license, unless indicated otherwise in a credit line to the material. If material is not included in the chapter's Creative Commons license and your intended use is not permitted by statutory regulation or exceeds the permitted use, you will need to obtain permission directly from the copyright holder.

2 Incremental, Iterative, Transformative: A Social-Learning Approach in Spatial and Transdisciplinary Research and Practice

Ignacio Castillo Ulloa

2.1 Seeing Spatial and Transdisciplinary Research Through 'Interrelated Strings' of Social Learning

In this chapter, I explore how research and practice that draws on spatial and transdisciplinary methods and approaches can be enhanced or even hindered by the notion of *social learning*. After having provided a working definition of social learning; the distinction between multi-, inter- and transdisciplinarity; and my view on spatial and transdisciplinary research and practice, studies on social learning relative to planning (von Schönfeld et al., 2020), transdisciplinary research (Herrero et al., 2019), transdisciplinary co-production (Slater & Robinson, 2020) and action-oriented knowledge (Caniglia et al., 2021) are reviewed. Given that these debates urge for a change of paradigm regarding the way sustainability is to be characterised and pursued, potentialities as well as challenges to render novel attempts at sustainability *transformative* are identified. Afterwards, I briefly reflect retrospectively on my own practical experience participating in the development of master plans in Costa Rica, my homeland, in the light of the threefold conceptual prelude—social learning; multi-, inter- and transdisciplinarity; and spatial and transdisciplinary research—and some reflections extracted from the previously mentioned studies. In doing so I conceive transdisciplinarity and spatial methods being both understood and rendered operative according to rather different, at times even disparate, logics within research and practice. By and large, I conclude that more pragmatic ways to promote sustainable cities and communities need not only champion social learning, to then make the most of spatial and transdisciplinary research and practice, but also, besides transformative, practical-normative implementations ought to be both *iterative* and *incremental* as well as aimed at systemic change.

2.2 Delineating the Theoretical Purview

2.2.1 Circumventing Ambiguous Rhetoric: Definitional Boundaries of Social Learning

Social learning, loosely speaking, requires at least two basic elements to materialise: the involvement of two or more individuals or groups of individuals and some sort of interaction between them. Moreover, these interactions are

---

I. Castillo Ulloa (✉)
Department of Urban and Regional Planning + SMUS, Technische Universität Berlin, Berlin, Germany
e-mail: i.castilloulloa@tu-berlin.de

connected in different ways and to varying degrees with knowledge (its creation, development, reproduction, validation,…) and distinctive aims. Hence, social learning is both *interactive* and *purposive* in character. Seen like this, the *learning* dimension confers a normative value, for social learning, irrespective of the goal being pursued, is de facto an end in and of itself (Pahl-Wostl, 2006). In addition, with its conceptual roots in the realm of psychology and first coined by Bandura (1971), social learning is thought to allow for the necessary human interactions to be performed for novel knowledge to be produced and through which adaptation and responses to complex challenges may be obtained (Wells, 2012; Folke et al., 2003). Next to its *substantive* (the inherent learning formative factor) and *normative* (the facilitation of goal achievement) beneficial features, social learning can also be *instrumental* (Slater & Robinson, 2020: 3). As various authors point out, the instrumentality of social learning becomes apparent when, for example, it increases trust, furthers governance, encourages changes in attitude and behaviour, fosters social legitimacy and empowers stakeholders and networks (Beers et al., 2016; Popa et al., 2017). Similarly, when social learning gives way to new and more socially and ecologically robust identities, individual and collective capacities and institutions, social learning constitutes a substantive, normative and instrumental process (Pahl-Wostl, 2006).

These arguably objective and well-defined features of social learning notwithstanding, the term is still treated with relative ambiguity and even abandon in the literature. As Parson and Clark (1995: 429) observe:

> The term social learning conceals great diversity. That many researchers describe the phenomena they are examining as 'social learning' does not necessarily indicate a common theoretical perspective, disciplinary heritage, or even language. Rather, the contributions employ the language, concepts and research methods of a half-dozen major disciplines; focus on individuals, groups, formal organizations, professional communities, or entire societies; and use divergent definitions of learning, of what it means for learning to be 'social', and of theory.

Acknowledging this caveat, and for the sake of conceptual clarity and consistency throughout my argumentation and to render its definition operative, social learning is deemed to be:

> The process that allows individuals and groups of individuals from creating to developing to exchanging (experiential, abstract, practical,…) knowledge through varied types of dialogic interactions. Whether knowledge is produced, developed or exchanged depends on the circumstances under which the interactivity underpinning the social learning unfolds (distribution of resources, power asymmetries, objectives sought,…). Likewise, according to its substantive, normative or instrumental components, social learning furthers change by new knowledge production just as nurtures resistant to change by corroborating or reassessing existing knowledge. By the same token, social learning characterises knowledge: it makes tacit knowledge visible, practical knowledge tactical and strategic, abstract knowledge plural,… Finally, social learning, as mutual cognition, is both individual and collective and, rather than finite, social learning instances are inscribed in continuing strings of interrelated social learning processes.

Stretching the line of conceptual legibility and evenness, in the ensuing subsection I sketch out my view on research and practice that have a spatial and transdisciplinary character.

2.3 Sharpening the Focal Take: Delineating the Spatiality and Transdisciplinarity of Research and Practice

2.3.1 Untying Multi-, Inter- and Transdisciplinarity Knots

Let us begin with the Gordian knot that supposes telling multi-, inter- and transdisciplinarity from one another in the field of research. The terms are sometimes used in literature interchangeably and even synonymously. However, there are nuanced, yet relevant, distinctions between what multi-, inter- and transdisciplinarity (might) mean. As shown in Fig. 2.1, out of transdisciplinary exchanges what can actually be deemed as a 'new discipline' (D_{new}) may emerge. Let's take as an example the case of biochemistry, whose origin

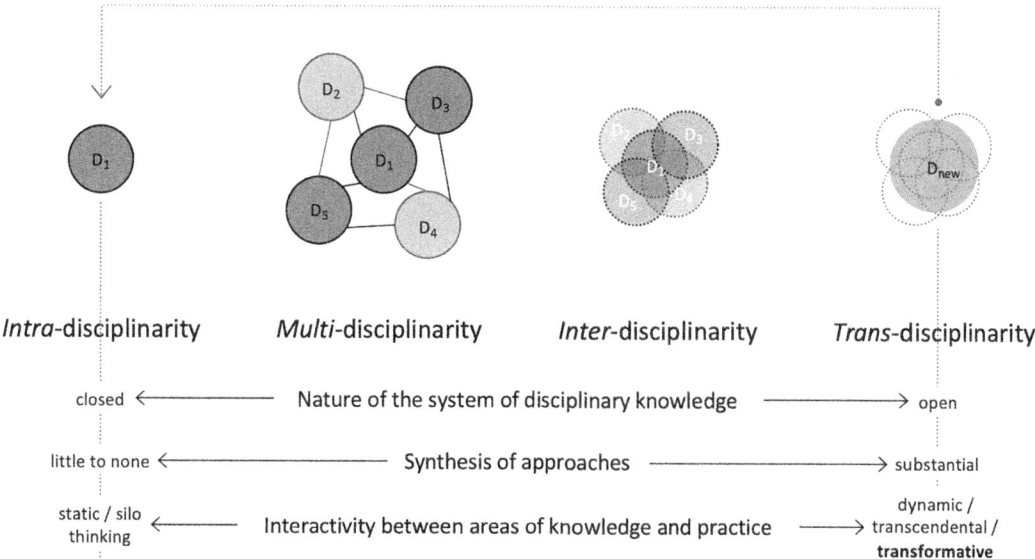

Fig. 2.1 Diagrammatic representation of the diverse types of interaction that may take place between disciplines. Note: the hyphen between the respective suffix and 'disciplinarity' is only meant for readability purposes and does not purport to be the proper spelling. (Source: Own elaboration based on Nicolescu (2002) and Appel and Kim-Appel (2018))

goes back to the disciplines of biology and chemistry overlapping to the point of their boundaries being no longer recognisable. Thus, transdisciplinary interactions may, but not exclusively, lead to intradisciplinarity. Here, it is important to observe that, while the diagram subtly alludes to a succession from intradisciplinarity to transdisciplinarity (and all the way back to intradisciplinarity), the linearity of representation of the types of interaction is only illustrative. It should therefore not be assumed that multidisciplinarity is somehow a prerequisite to interdisciplinarity, and interdisciplinarity in turn to transdisciplinarity. It may well be that, within a research or practice-led project (as will be discussed later), diverse forms of interaction transpire. Ergo: the interaction types do not rule each other out. Nor should any of them be considered better than the other. Instead, '[m]ultidisciplinarity, interdisciplinarity and transdisciplinarity are better treated as complementary rather than being mutually exclusive. It is important to stress this complementarity because without specialized disciplinary knowledge studies there would be no in-depth knowledge or data' (Lawrence, 2010: 21).

The differences between the diverse forms of disciplinary interaction can also be underscored in terms of: (i) the nature of the system of disciplinary knowledge, (ii) the synthesis of approaches and (iii) interactivity between areas of knowledge (see lower section of Fig. 2.1). Jonathan Appel and Dohee Kim-Appel (2018: 63ff.), following Nicolescu (2002), claim that under intradisciplinary circumstances at play is a 'closed system' for only one single discipline (D_1) is at work. Thus, there is essentially no synthesis of approaches. Furthermore, there is a fixed interactivity between areas of knowledge for there is just one. Multidisciplinary comes into existence, continue the authors, when researchers from diverse disciplines (D_1, D_2, D_3,…) enter an interactive arena to work together by making use of their own disciplinary knowledge. Characterised by a still rather sealed system of disciplinary knowledge (researchers pretty much stay in their corners), the synthesis of approaches is somewhat partial, insofar as the input of each discipline is still discernible. Hence, while there is already interactivity happening, it 'travels' from one discipline to the other.

Moreover, in interdisciplinarity, the system of disciplinary knowledge broadens significantly because the boundaries become porous enough to allow 'integrating knowledge and methods from different disciplines, using a real synthesis of approaches – but still disciplinary' (Appel & Kim-Appel, 2018: 64). Lastly, transdisciplinarity occurs to the extent that a system is absolutely open, thereby 'creating a unity of intellectual frameworks beyond the disciplinary perspectives' and 'new "transcendent" areas of knowledge and practice' (Ibid). In an akin manner, Ramadier (2004: 425) sees transdisciplinarity interweaving disciplines, rather than simply articulating types of disciplinary knowledges, which is achieved via multi- and interdisciplinarity. Consequently, while multi- and interdisciplinarity are concerned with *knowledge unity*, transdisciplinarity finds expression in *knowledge coherence*.

Following more closely Nicolescuian transdisciplinarity and as sustained by Appel and Kim-Appel (2018: 62), I see transdisciplinarity as neither a novel discipline nor a 'super-discipline'. Rather, transdisciplinarity is the interactive and transformative dynamics of research that are nurtured by various open systems of disciplinary knowledge whose inputs are clarified and refined by the complementarity that characterises *transdisciplinary coherent knowledge*. That being so, a critical question emerges: How is such knowledge set into practice or what does a transdisciplinary practice look like? I next explore this.

2.3.2 Transdisciplinarity in Practice—An Importance Nuance

With the advent of the Anthropocene era, trends of societal, economic and environmental changes have been hastily and unevenly accentuated in cities and regions throughout the world. A myriad of closely interlinked effects and aftereffects—loss of biodiversity; air, water and soil pollution; shortages of food; and overconsumption of resources—present scientists and policymakers alike with pressing and ostensibly unsurmountable challenges. The peak of these challenges has translated into the so-called wicked problems (Rittel & Webber, 1973) 'which are so complex and interconnected that they cannot really be *solved*, rather only *resolved* in multiple ways, with differing costs and benefits for those involved' (Lawrence et al., 2022: 44; *italics* in the original).

Diverse attempts at understanding and reframing the nexus between science and policy, in such a way that it helps to resolve wicked problems substantively, have been pursued through the long-dated development of transdisciplinary research (Jantsch, 1970, Hadorn et al., 2008, Bernstein, 2015[1]). In time, transdisciplinary research derived into various perspectives and approaches that, while meant to underlie sustainability-oriented measures and policies by, for instance, involving societal actors who would eventually be affected by them, do not always include societal actors into the research process (Lawrence et al., 2022: 45). Moreover, policy advice has conventionally and all too often focused on looking into and highlighting the consequences, either desired or otherwise, of already implemented policies and projects. Although this approach may well work for singular, non-ambiguous and simplistic instances of science-based policy (e.g. location of measuring instruments to collect data to inform regulations), it falls short when it comes to multifaceted phenomena characterised by the constant variability and interdependence of a wide array of (social, economic, political, environmental,...) factors such as the currently heightening climate change crisis. Here is where transdisciplinarity begins to move out of the realm of research and enters that of practice inasmuch as *transformative transdisciplinary knowledge* is very much required to inform policies that could effectively begin to mitigate and eventually render manageable the

---

[1] See for a thorough reconstruction of the genealogy of transdisciplinarity.

impacts of climate change. In this regard, it is worth noting that 'transdisciplinary practice' goes well beyond 'purely research activities' (formulating hypothesis, designing a research method, deploying research tools to collect data, and the like) and includes what can be seen as 'transdisciplinary' meetings, working groups and agendas, in which non-academic societal actors, practitioners and researchers from diverse disciplines converge. At least in principle, this 'transdisciplinary actors' collaboration' is intended to enhance, rather than substitute, both the use of longstanding scientific methods and traditional policy advice formats (Lawrence et al., 2022: 45). Nevertheless, the reality of these partnerships may turn out to be rather different. Now, I move on to address the last element that outlines the conceptual take: spatial methods and spatial research.

2.3.3 Methods Are Not Spatial but Rather Enable to Fathom Out Phenomena Spatially

What characterises methods as spatial is not some sort of intrinsic and pure spatial nature. Rather, distinctive methods can be calibrated to either quantitatively or qualitatively (or both) shed light on the *spatial dimensions* undergirding the empirical phenomena of interest. What those spatial dimensions are and how they are to be approached depends on other inquiry components such as research questions, conceptual framework, goals, time frames, etc. Spatial research, a term that has come progressively into vogue, seems to encompass the many possible forms and purposes that examining empirical phenomena spatially may take and have—including even disciplinary intersections. With an across-the-board impetus, spatial research has positioned itself at the forefront of research in various fields: geography, social sciences and humanities, architecture, urban design, planning and a host of others. Each championing their traditions, methods and tools of inquiry, there seems to be little to no consensus among disciplinary approaches as to what spatial research could ultimately be like. Thus, the designation of either methods or research as spatial must be taken with caution for it may well end up being mere euphony. This observation is vital if spatial research is to be coupled with transdisciplinarity.

In this regard, the distinction between 'graphical means of representation' (that is, tangible outcomes such as maps, diagrams, sketches, videos, pictures,…) and the actual material tools employed to produce visualisations (from software like GIS to cameras to paper and pencil) holds great relevance. For starters, it must be acknowledged that graphical elements, far from only being useful to prepare and present research results or constituting themselves research outputs, are integral to, for they are instrumental in, both collecting and analysing data. The implication here is that the tools and the modes that are deployed ought to have a certain degree of 'neutrality' to the extent that spatial research is to be transdisciplinary (a condition that also applies to practice-oriented projects). To wit, neither any form of visualisations nor the way to produce them must be subjected to a 'disciplinary canonisation' that prevents researchers from other fields to make 'adequate' use of them. To illustrate the significance of this point, let us consider, as an analogous example, what has happened with the qualifier 'ethnographic' which, as Tim Ingold (2014: 384; *italics* in the original) sharply contends, is unwarily applied everywhere: 'there is ethnographic encounter, ethnographic fieldwork, ethnographic method, ethnographic knowledge. There are ethnographic monographs, and ethnographic films. And now we have ethnographic theory! Through all these runs the ethnographer. Taking this as a primary dimension of identity, it would appear that everything the ethnographer turns his or her hand to is, *prima facie*, ethnographic'.

This intrinsic capacity—i.e. that everything an ethnographer does is rhetorically ethnographic— 'magically' bestowed upon the ethnographer is, in an akin manner, perceptible when, for instance, it is assessed that: only cartographers can properly create maps, architects pencil sketches, social scientists conduct qualitative interviews, and so forth. What is ultimately at issue is that such approach and line of reasoning, in view of furthering interactions between disciplines to elucidate phenomena empirically and spatially, will at best lead to multidisciplinarity and at worst to intradisciplinarity—with its silo thinking and closed systems of knowledge. To circumvent this trap, social learning may indeed serve as a catalyser that aids to move beyond 'canonical intradisciplinarity'. It would then be possible to not only look into phenomena empirically in a substantive transdisciplinary manner but also fathom them spatially out. To explore how this may or may not happen within a specific practical case, I next discuss how social learning is understood and what role it plays in specific knowledge domains and research circumstances.

2.4 Telescoping the Gist of Social Learning's Instrumentalising within Specific Instances of Research and Practice

In this section, I draw on specific sources to underscore how social learning is viewed and rendered instrumental in connection with *planning processes* (von Schönfeld et al., 2020), *transdisciplinary research* (Herrero et al., 2019), *transdisciplinary co-production* (Slater & Robinson, 2020) and *action-oriented knowledge* (Caniglia et al., 2021). The selection of these particular studies was done considering their common assertion that social learning has a lot to offer for the promotion of urban sustainability. On such account, it should be noted that the discussion of the potentials social learning can represent a spatial and transdisciplinary research and practice cannot be circumscribed to one or two targets of SDG11. Not only may these potentials well apply to other SDGs (let alone that the collection of SDGs conform an interconnecting system[2]), but also the very character of social learning—seen as a process whereby knowledge (from skills to experiences) is developed and exchanged—confers it a sort of universal applicability.

2.4.1 Social Learning as a Planning Practice and Urban Development Catalyser

Existing literature makes apparent both the connection between urbanisation and planning processes and how this interplay, when fuelled by social learning, furthers actual development (as opposed to sheer growth) (Rydin, 2010; Scholz et al., 2014; Young, 2009). This, however, is contingent not only on the particular understanding of social learning at play but also on how both planners and non-planners interacting with one another instrumentalise it. Planning studies tend to frame social learning through overtly positive narratives that somewhat uncritically equate it to the achievement of desirable outcomes (Dumitru et al., 2017; Holden et al., 2014). As von Schönfeld et al. (2020: 412) contend:

> Social learning as studied in planning and related disciplines is often entangled in three interpretations. First, social learning is understood as an inevitable process resulting from interaction among actors. Second, social learning is seen as an agenda that should be embedded in planning practice. Third, social learning is considered as a process that intrinsically leads to desirable and constructive outcomes.

By focusing on the individual level of learning, given the relevance of personal dynamics entangled in planning practice (Tewdwr-Jones, 2002), these authors map moments of social learning within planning processes by way of three driving questions —'who learns', 'what' and 'from whom'—to demonstrate that social learning can lead to desirable and undesirable

---

[2] Griggs et al. (2013: 307) explain, about this aspect of the SDGs, that the 'SDG framework manages trade-offs and maximises synergies between targets, and can be implemented from international to city scales. It integrates social, economic and environmental dimensions and provides guidance for humanity to prosper in the long term'. Similarly, Folke et al. (2016: 41) illustrate how the SDGs are interrelated along a threefold axis that goes from the biosphere to the economy level.

consequences alike.[3] Building on the analysis of a case study explored through storytelling, von Schönfeld et al. (2020: 428) arrive at the conclusion that practicing planning in a one-sided manner, thereby downplaying (if not outright excluding) the knowledge other actors embody, is no longer viable to the extent that social learning, as an inherent element of planning practice, is meant to give way to results that are beneficial to all parties involved. More concretely, their findings underscore the need for planners to 'be mindful of individual backgrounds and motivations, and how combinations of these can lead to a large variety of outcomes in terms of the planning and [social] learning outcomes' (von Schönfeld et al., 2020: 428). Otherwise, the chances for social learning to catalyse planning practice and urban development are little to none.

2.4.2 Rendering Transdisciplinarity Transformative

Despite the lack of consensus on what transdisciplinarity research may ultimately be, a number of scholars (Jahn, 2008; Popa et al., 2015; Bieluch et al., 2017; Plummer et al., 2022) agree on how the various challenges that furthering sustainable urban development betokens can be substantially addressed through transdisciplinarity, for it allows to deal more properly with the inherently complex socio-ecological interdependencies. This, however, begs the question of how diverse epistemic, social-organisational and communicative traditions may be reconciled to, subsequently, allow for novel knowledge to arise, societal and scientific problems to relate, etc. As a dimension that transdisciplinarity, as an approach, entails, social learning contributes to this purpose, thereby setting the stage for potential problem-solving. Because social learning enables scientific and non-scientific actors to reassess their deep-seated background assumptions and values, a much-needed common understanding of not only the context but also any potential (re)solution of the problem at stake can be then achieved (Hadorn et al., 2006; Wickson et al., 2006). However, while collaborations between scientists and practitioners may indeed come to fruition by way of transdisciplinarity research, participants usually mobilise and deploy the tools and methods of their respective disciplinary fields without barely considering, if at all, the impacts this have on the social learning process they are embedded in (Hegger et al., 2012; Lang et al., 2012). As a way out this conundrum, Herrero et al. (2019: 752ff.) outline a strategy consisting of a series of 'key process features' (e.g. openness of the co-construction process, clarifying the normative background, the existence of an authentic rational dialogue, among others) that should become operative at two distinctive levels:

1. Social learning through mobilising existing reflexivity within the process, also called *deliberative reflexivity*, which relates to how much the understanding of the practical problem situation and the formulation of the research question have been discussed by all the actors and called into question through argumentation on the values, the epistemic criteria and societal objectives.
2. Social learning through generating and transforming reflexive capacities, also called *pragmatist reflexivity*, which refers to how much the practical process of collaborative problem solving and experimentation have created new actor competences and built new capacities to critically assess values and the understanding of the practical problem situation (Herrero et al., 2019: 752; *italics* in the original).

Against this backdrop, transdisciplinarity, as a research approach, becomes in effect transformative to the degree that social learning, as the bedrock of collaborative transdisciplinary processes, complies with at least three conditions (which in turn attest to its occurrence):

---
[3] The (degree of) desirability of the consequences that social learning may cause through planning decisions and actions cannot be unequivocally determined. This is highly dependent on, among other factors, the distribution and relations of power among the participants, which, in the case of planning and concomitant social learning processes, is for the most part uneven and relatively stable.

(i) All individuals partaking in the social learning process have undergone a change in understanding; that is, learning on both cognitive (value- and epistemology-laden conventions) and practical (e.g. reassessment of the situational character of the research-practical problem) aspects.

(ii) This change has gradually moved from each individual to wider communities of practice, thereby scaling up the learning range for inquiry and practical challenges to be coupled.

(iii) The now widespread change emerges from ideas, proposals, information, know-how, arguments, and the like being swapped between larger social units and communities of practice based on a mode of a shared communicative rationality. (Reed et al., 2010, Herrero et al., 2019: 753–754)

Given this threefold change social learning propels by dint of *deliberative* (closely related to the first 'change fold') and *pragmatist* (buttressing the second and third 'change folds') reflexivity, the transformative potential of transdisciplinarity develops 'by a combination of co-construction methods that explicitly address normative agendas and orientations, and appropriate governance of power relations amongst social actors and scientists' (Herrero et al., 2019: 752). Consequently, 'transdisciplinary research projects may contribute to generating effective social learning on sustainability issues' (Ibid).

2.4.3 Social Learning as the Building Block of Sustainability Transitions

In the face of the various wicked problems that strike worldwide, transitional scholars advocate for incremental improvements to be replaced with fundamental systemic transformations that could lead the way to attaining sustainability (Markard et al., 2012; Sol et al., 2018). These systemic transformations, moreover, would reshape technologies, institutions, practices, polices and mode of thinking through social learning and its ensuing alternative knowledge production processes (Kemp et al., 2009; Beers et al., 2016). Slater and Robinson (2020: 3) see the need for social learning, in view of its ingrained conceptual ambiguity, to be grounded in real-life circumstances such as transdisciplinary knowledge coproduction (to then trigger sustainability-oriented transitions).

Within this kind of contextual settings, the authors argue, both material and immaterial components determine whether and the extent to which social learning occurs. And, for the case of transdisciplinary knowledge coproduction, when seen through a social practice framework, it comprises *materials* (infrastructures, tools, equipment and even the body itself), *skills* (know-how, dexterities, assessments and the like that are acquired, enacted and developed further through bodily performance) and *meanings* (socially shared concepts, constructs and ideas that render symbolic and purposeful the participatory practice) (Slater & Robinson, 2020: 3–4, Shove et al., 2012: 22–25).

Examining through the materials-skills-meanings social practice lens what is identified as a transdisciplinary knowledge coproduction effort between researchers based at the University of Toronto, representatives of The Atmospheric Fund (TAF) and the City of Toronto and neighbourhood leaders, Slater and Robinson (2020: 7) deconstruct the social learning process into three analytical tiers—processes, outcomes and impacts[4]—to state if social learning upshots 'coalesce around (i) new knowledge (and knowledge products), (ii) new actions and/or behaviours and (iii) new relationships' (Ibid: 6). In their discussion of the case study, the authors emphasise how 'plural forms of social learning (processes and outcomes)' were generated through the direct engagement of participants with a diversity of social practice elements at hand and by enacting their own competences and rendering therewith the collaborative effort meaningful. In brief, the tripartite analytical take developed to unravel the dynamics of

---

[4] For each tier, a series of principles and criteria were also developed (Slater & Robinson, 2020: 7).

transdisciplinary knowledge coproduction makes noticeable how social learning is not to be deemed as finite, but rather, as previously argued, strings of continuous and ever-evolving social learning processes that allow for novelty to arise. Insofar as social learning gives way to new actions, relationships and knowledges, which expand to broader social units, as well as discursive spaces 'for the exploration and deep engagement with contested notions of sustainability, entwined social learning and [transdisciplinary knowledge] coproduction efforts operate as a powerful means of responding to sustainability challenges and conceiving of new pathways for transition' (Slater & Robinson, 2020: 17).

2.4.4 Overcoming Technocracy and Furthering Sustainability Through Action-Oriented Knowledge

The crucial importance of system-wide and transformative change to further sustainability, as various scientists have urged (Díaz et al., 2019), rests on research that brings together policymakers (working at the national, regional or local level), civil society (in particular, traditionally underrepresented minorities and organised communities) and entrepreneurs. As a result, as Caniglia et al. (2021: 93) emphasise, this 'aspiration also demands that the research community engages more actively with action-oriented knowledge for sustainability'. This call echoes a trend of scholarship that has been pushing for different (and even rather counterintuitive) ways to do research as well as to bridge the gap between science and society at large (Lubchenco, 1998; Nowotny et al., 2001; Ravetz, 1999). Labelled as 'Mode 2' and 'post-normal' science, Caniglia et al. (2021: 94) identify, as their common denominator, the reframing of 'processes of knowledge production in our contemporary world as socially distributed and subject to multiple accountabilities' and regard them as the pillars of 'action-oriented research in sustainability'. Through a review of a series of studies in sustainability science that combine knowledge, action and capacity building, Caniglia et al. (2021: 95) propose, using a philosophical framework that distinguishes action from behaviour and habits, that actions for sustainability ought to be: 'intentionally designed to create transformative change towards sustainability; [...] involve shared agency of multiple actors; and [...] materialize through contextual realization in constantly evolving and emergent settings'. Along these three axes of action, the authors' assessment traverses a great variety of knowledges that informs *intentional design* (generative, prescriptive, strategic knowledge), enhances *shared agency* (critical, empowering, co-produced knowledge) and enables *contextual realisation* (emergent, tactical, situated knowledge) (Ibid: 96). And, while acknowledging that this diversity of knowledge types is not meant to be exhaustive, Caniglia et al. (2021: 97) emphasise that individual and social learning, as the underpinnings of the sought after co-production of action-oriented knowledge, 'require that societal and academic actors develop a capacity for knowledge pluralism: the ability to appreciate and work with multiple kinds of knowledge'. Only then, it is asserted by the scholars, definitive technocratic solutions to sustainability can be downright rejected, as the many kinds of knowledge, dovetailed by way of individual and social learning, give way to the shared design, enactment and ultimate realisation of change.

2.4.5 Social Learning Is Not the Bullseye—But Rather What it Enables

Each of these four reviewed studies, both in its own right and rather inadvertently in connection with one another, shed light on the manifold potentials that social learning carries to further transdisciplinarity, collaborative forms of knowledge creation and coordinate anew efforts between science and society. At the same time, both caveats (e.g. the latency of undesirable outcomes) and shortcomings (e.g. issues pertaining the reassessment of individual values and collective conventions) of social learning are acknowledged (for the most part tangentially though). Although on the surface it may appear that social learning is at the crux of the matter, the purpose of this section was to highlight the instrumental—and not the conceptual-definitional—dimension of social learning. Hence, beyond the particularities that delimit the discussion developed in each study either around or in relation to social learning, the emphasis, at the end of day, is placed on its indelible *enabling* nature.

Having said that, there are a few reflections to highlight. For example, there is no contemplation about the distinction between transdisciplinarity in research and in practice. It is instead treated as a given that transdisciplinarity is always *simply happening* (let alone the possibility that, rather than transdisciplinarity, it might be multi- or interdisciplinarity what is transpiring). Similarly, while proposals like considering extra-scientific insight and knowledge as an integral part of transdisciplinarity are both provocative and valuable, it is not that clear how *extra-scientific* and *discipline* may actually constitute two sides of the same coin (which, in any case, does not at all diminish the pertinence of integrating forms of knowledge production that do without institutional sanctioning into social learning processes). Likewise, although research methods do not go unnoticed their specificities, particularly if they are used to look into the problematic challenges standing on the way to sustainability from a spatial angle, are left aside. Nor is any particular attention paid to how research tools and methods shape and are shaped through the interactions of the participants involved in the diverse strings of social learning processes. With that in mind, in the next section, I revisit my practical experience participating in the creation of master plans.

2.5 Will-o'-the-wisp: The Pieces May Be at Hand but Won't Necessarily Fit

Master plans lie at the core of official-formal planning practice in Costa Rica and are developed for or on behalf of the state to foster *ordenamiento territorial* (roughly, 'territorial ordering'). Moreover, the arrangement of the territory through master plans is furthered at diverse physical scales that do not entirely tally with the official territorial division of the country.[5] At least in principle, the master plans constitute the outcome of a collective and collaborative endeavour between professionals from diverse disciplinary backgrounds: planning, civil engineering, geography, topography, cartography, hydrology, biology, environmental science, architecture, anthropology, social work, law, economics and a host of others. In addition to that, this master planning practice is not only remarkedly state-led but also, due to the effect of planning ideas derived (and somewhat adapted) from the American planning tradition, accentually statutory. Between 2004 and 2007, I worked as a member of a group of researchers-practitioners[6] producing *planes reguladores*: local master plans envisioned to become operative at the geographical and politico-administrative level of cantons (see Note 5).

The municipalities (i.e. the local governments) are to formulate the *planes reguladores* in partnership with representatives of the *Instituto Nacional de Vivienda y Urbanismo* (INVU, the National Institute of Housing and Urbanism). Alternatively, and should they have funds at their disposal (e.g. donations by NGOs and foundations or direct funds transfers from the central government), municipalities can commission the master plans to private firms or public universities. Either way, the municipal authorities must oversee the production of and, once completed, enact the *plan regulador*. To ensure consistency throughout the generation process of the *planes reguladores*, the *Dirección de Planificación Urbana* (Urban Planning Directorate) of the INVU together with the *Ministerio de Vivienda y Asentamientos Humanos* (Ministry of Housing and Human Settlements) and the Project

---

[5] It is as follows: provinces, cantons, districts and barrios.
[6] The hyphenation denotes here how blurred the line that divided our roles as either 'researchers' or 'practitioners' actually was.

Management Unity of the PRU-GAM Project[7] put together a handbook. The *planes reguladores* are in this document defined as a 'technical instrument' whose objective 'is to create a structural model, balanced, efficient, hierarchical and in complete harmony with the environment and the national idiosyncrasy [...] to generate more human cities, in keeping with urban work, beautiful and with an improved quality of life' (INVU, 2006: 1; own translation). Moreover, the manual lays out a ubiquitous procedure to create the *planes reguladores* that consists of six consecutive phases: (1) data collection, (2) analysis and diagnosis, (3) forecast, (4) proposals, (5) approval and adoption and, finally, (6) management for implementation (INVU 2006: 9).

This mode of master planning is ostensibly quite exhaustive. Accurateness is given a prominent relevance throughout these six steps, for the data gathering, systematisation and analysis are meant to be comprehensive, results obtained accurate and intricately portrayed, and both scenarios and proposals definitive. However, the practical implementation of such planning method, as I argued elsewhere (Castillo Ulloa, 2016: 25), remains still rather basic and turns out to be quite problematic, given the imperative and prevalent character of the theoretical framework wherefrom it originated, which visibly resonates with the principles of the rational-comprehensive planning paradigm at whose core appears the figure of the 'planner-analyst' guided by the dictum of 'the more comprehensive the analysis of the planning problem[s], the better the plan' (Mäntysalo, 2005: 24).

Against this backdrop, in what follows, I focus on the experience I had contributing to develop *planes reguladores* during the first two phases (data collection and analysis and diagnosis) in particular and place emphasis on the usage of maps to illustrate some of the aspects elaborated upon in the previous sections: for instance, how the interaction between the assorted disciplines was, how research methods were used to look into phenomena spatially and how the strings of social learning were unfolding.

2.6 Partition and Addition: The Underlying Logic of Master Planning Through *Planes Reguladores*

During the time I was actively contributing to the production of *planes reguladores*, the data gathering and its subsequent systematisation, analysis and assessment were allocated following a rationale that linked the researcher-practitioner and their respective disciplinary background with specific subject matters (e.g. solid waste management, transportation and road networks, neighbourhoods and urbanisation patterns, weather and natural disasters, economic activities and employment and a host of others). Although the staff comprised several disciplines (with a rather dominant presence of civil engineers), it was customary to outsource studies on specific subjects like hydrology, history and biology to freelance consultants. This practice even further accentuated the separation between disciplines involved in the process, as meetings to discuss with hired consultants their progress or deliverables were hardly ever convened (feedback was usually provided in written form by one or two team members at most). After each researcher-practitioner of the team (sometimes working in tandem or small groups of the same or similar discipline) had received their task, data began to be compiled thorough various means: collecting statistics, on-site data survey, literature research, review of existing studies and the like. In my case, the vast majority of the data I had to gather was through fieldwork conducting interviews with residents and local officials, meetings with local authorities, visual documentation (taking photos and, at times, sketching) and georeferencing the location of the surveyed barrios, towns and villages throughout the canton for which the *plan regulador* was meant. Over a span of 10–12 months, a series of thematic and discipline-specific assessments were completed and

---
[7]This project was aimed at generating a new regional plan for the *Gran Área Metropolitana* (GAM), the country's largest urban agglomeration, with financial aid provided by the European Union.

eventually delivered in predominantly textual form and supplemented by charts, diagrams, pictures and, particularly, maps. Once all contributions were ready, they were aggregated to give way to a lengthy document (up to 500 pages or more) called *diagnostic study*—which constituted the first of a series of outcomes that needed to be submitted to the municipality that had commissioned the *plan regulador*.

As shown in Fig. 2.1, the logic of work to produce this study was, for the most part, intradisciplinary, given the way every assessment per topic was produced: basically in 'disciplinary isolation'. Nevertheless, since the aggregative way in which research outputs per topic were compiled did require some interrelation (but not to the point of overlapping) between disciplines, a degree of multidisciplinary was also occurring. Oddly enough, the little level of interaction across disciplines, rather than by chance, was by design: the evident compartmentalisation was meant as a coping mechanism of budget and time constrains. Furthermore, there were arguably 'temporal instances of interdisciplinarity' taking place during internal team meetings to discuss, among other aspects, hurdles encountered (e.g. quality of or inaccessibility to data) and share methodological pointers to deal with them. Given that interdisciplinarity, as sustained by Nicolescu (2014: 9), 'concerns the transfer of methods from one discipline to another', but all the while remaining 'within the framework of disciplinary research', by way of methodological suggestions a certain level of interdisciplinary interaction, albeit low and quite specific, was in effect happening. Likewise, and admittedly being a bit of a stretch, if transdisciplinarity is taken to be Mode 2 research (see Note 2), such deliberations can be deemed to be transdisciplinary insofar as a group of research-practitioners came together for short periods of time to find concrete solutions. To be sure, social learning was very much not only embedded in the seemingly intradisciplinary production of the assessments per topic that made up the diagnostic study but also in the inter- and transdisciplinary team meetings that took place in parallel. As to the former, I next revisit my experience preparing the assessment about green open-space areas and, specifically, the preparation of maps to present the findings.

2.7 Muddling Through Fixed Roles and Duties: Attempting to Surpass Mapping's Inherent Reductionism

Mapping is by nature reductionist. For maps to capture and portray the essence of *that* which is being graphically represented, a process of oversimplification must occur: from cartographic maps seeking to fix objectively reality 'as it is' to mind/mental maps (seen as a mapping transposition) portraying subjective and even affective imaginations of reality 'as perceived'.[8] Now, with this, I do not imply that maps are somehow irredeemably faulty. But rather that considering maps as seamless, objective and legible representations of reality is not that practical and fruitful for, among other things, master planning purposes. As previously mentioned, maps were used as much as possible to amplify, by diversifying, the different means whereby the results included in the diagnostic study were per theme depicted. Amid various topics, I was in charge of researching green open-space areas, which ranged from parks to football fields to riverbanks. To start off fieldwork, I would rely on the municipality's inventory and, if available, map of the official (i.e. those which were owned or administered by the local government) green open-space areas to plan in situ data gathering. The data collected usually encompassed pictures, sketches, fieldnotes and, if needed, the geo-localisation of the surveyed green open-space areas. In addition, I interviewed nearby residents and persons in charge of, for instance, opening and closing the area in case it were fenced in. The analysis also

---

[8] Mapping and, by extension, maps are a much more complex and vaster subject of discussion that goes well beyond the boundaries of this chapter and subsection. For the sake of clarity, what I describe here is a succinct reflection of the kind of maps and mapping I was engaged with while producing the *planes reguladores*: cartographic maps.

included non-official green open-space areas that were spotted during fieldwork: for example, if I noticed that people appropriated a vacant lot temporarily or I was told during interviews with residents or local officials that a section within a private nature reserve is open to the public (sometimes at a modest entry fee). In so doing I was conceptually deviating from the rather narrow definition of green open-space areas municipalities stated officially, namely, those under their jurisdiction. Likewise, one of my objectives was to typologise the analysed green open-space areas, so that, on the one hand, their management could be tailored accordingly and, on the other hand, specific proposals (e.g. buying lots with potential to become or expand existing green open-space areas) could later be developed. Moreover, prior to concluding the input for the diagnostic study, I usually presented this strategy to and reviewed it with the local officials at the municipality responsible for the management of green open-space areas, who recurrently expressed mix reactions. While, at times, the idea of incorporating non-official green open-space areas to the existing inventory was welcome, obstacles and implications were afterwards listed without any effort to propose ways to circumvent them. Hence, dialogic efforts to take the issue a step further ended up, more often than not, in a logjam.

Once the data compilation phase was completed, I would begin to draw up a draft of the assessment report—that is, my 'disciplinary input' for the diagnostic study. To this end, a series of maps were put together: from a general map indicating the location of each surveyed green open-space area to a map indicating the ratio of available square metres of green open-space areas per inhabitant. To create these maps, I had to cooperate with a team colleague who, for all intents and purposes, acted as the 'sanctioned cartographer' on the grounds that they possessed the technical know-how required to generate the maps, namely, the use of ArcGIS software. Although the process of turning the gathered data into information by dint of the maps (to eventually give way to the epistemic basis that buttressed the policies and norms of the *plan*

regulador), in my view, must have been more dynamic and interactional, the fact that specific roles and tasks were assigned actually deterred any chances for it to happen. Therefore, constant frictions and misunderstandings encumbered the social learning that should have resulted in a more interdisciplinary production of maps. A case in point is what transpired with regard to the above-mentioned map showing the rate of square metres of green open-space areas per inhabitant. My intention was to factor in qualitative data obtained through interviews and photo documentation. Consequently, the transformation of such textual and visual qualitative data into graphic information, and its subsequent incorporation into its representation (i.e. the map), would have enriched the indicator of square metres per inhabitant by rendering visible visual components containing key pieces of information that would otherwise have been occluded. After all, graphical elements, as stated before, are very much part of both data collection and analysis. However, my then colleague repeatedly rejected this suggestion by alleging that information represented in maps had to stick to quantitative rigid parameters and rely solely on measurable data provided by the local government and public institutions or gathered through fieldwork (e.g. geopositioning). Doing otherwise would not only have been misleading and could have led to misinterpretations, but also gone against the established standardised procedure of map production that prevented qualitative data—provided it has not been rendered quantifiable beforehand—to be added in. Oddly enough, I felt that the sole expression of a ratio was indeed ambiguous, because it did not reflect 'other' realities about the availability and variability of green open-space areas I had discovered in the field: for instance, that some green open-space areas were not always accessible because they were fenced in, or that people actually took over green open-space areas not officially sanctioned as such. But be that as it may, the fact was that 'canonical interdisciplinarity' stood in the way of a more dynamic and ample portrayal of findings through maps.

All things considered, in hindsight, by both challenging the institutional convention of what a green open-space area was deemed to be and trying to extend the analysis beyond the somewhat platitudinous research question undergirding the analysis (i.e. what is the current state of the green open-space areas?), I was attempting to advance social learning, as expounded by Herrero et al. (2019: 752), through *deliberative* and then *pragmatist* reflexivity. As stated in the preceding section, according to the authors, process features like the openness of the co-construction of maps I was advocating for enable moving out of the deadlock that arises when researchers and practitioners partaking in a social learning process bring in their research methods, tools and idiosyncrasies without considering how this would play out. Nevertheless, since my former colleague and I—both being researcher-practitioners—could not work out our methodological differences and find common ground to maximise the assessment of the green open-space areas by surpassing the inherent reductionism of the maps, though deliberative reflexivity did transpire to a degree (we both did understand what the practical problem situation and research question at hand were), pragmatist reflexivity was palpably hindered.

2.8 Missing the Forest for the Trees: The Consequences of Disciplinary Overspecialisation and Task Compartmentalisation

Overall, the synergic continuity that could have emerged out of a more fluid and enhanced production of maps was hampered by the lack of neutrality that the tool employed to produce the maps, ArcGIS, actually had. Because only those regarded as 'competent enough' dictated the rules of how to use it, there were almost no chances to discuss further possibilities to integrate valuable elements derived from the experience of having gathered data upon which the creation of maps, which complemented the various thematic assessments, actually rested. Not only were details compiled in the form of field-notes and sketches, for example, de facto left out as a result, but the drafts of maps could also not be utilised as actual means to analyse the data prior to the production of their definitive version (i.e. those representing information). From the availability of green open-space areas to the morphology of the road system to the existence of potable water sources to natural risks latency, the analysis of the different topics comprised in the diagnostic study felt short within their respective disciplinary boundaries. Likewise, there was little chance to set the thematic assessments in relation with one another to then attenuate the strong multidisciplinary character of the diagnostic study—and, by extension, of the *planes reguladores*.

By the same token, another aspect notably hindering the possibility to infuse the production of the *planes reguladores* with at least a tinge of transdisciplinarity was the way non-scientific knowledge was regarded and treated. While citizen participation processes needed to be provided for and its outcomes incorporated, the input participating residents delivered was seen more as a desideratum than an actual asset. Thus, deliberative arenas, whereby a broader social learning between scientists, practitioners and residents of the canton for which the *plan regulador* was being created could be fostered, were kept from emerging. Nor could the epistemological base of the policies and norms that the *planes reguladores* contain be enhanced through the combination of scientific and extra-scientific insights. In other words, valuable non-scientific local knowledge and acumen, though not constituting sensu stricto disciplinary knowledge, were recurrently downplayed and even underestimated in their capacity to either improve or hinder the production of the *planes reguladores* (see Castillo Ulloa, 2016).

The way the production of the *planes reguladores* was tackled on the whole resembles that of putting together a puzzle through a linear and methodical aggregation, in view of the disciplinary overspecialisation and task compartmentalisation that permeated the entire process. Both intra- and multidisciplinary are attuned to this procedure but do not bode well for unravelling the full potential that *planes reguladores* (and master plans in general) have to offer for the promotion of more sustainable cities and communities. Such goal will continue to be elusive as long as having the pieces at hand (research methods and tools, researchers, financial resources,…) does not necessarily grant that they will fit remains unrealised.

2.9 From Talking the Talk to Walking the Walk: Some Reflections

Transdisciplinarity, neither in research nor in practice terms, can do without social learning; especially if transdisciplinarity is thought to nurture spatial research. Moreover, as I have made apparent throughout this chapter, there are no univocal conceptualisations or universal practical applications of social learning, transdisciplinarity and spatial research. Stark differences, nuances and similarities cutting across these three driving notions are important to acknowledge, according to the contextual circumstances in which they are rendered operative. Be it a research project carried out by a number of investigators from different disciplines, a practical project led collaboratively between scientists, practitioners and citizens or policies and norms for a master plan designed by a group of professionals with varied disciplinary training ideally drawing on non-scientific local knowledge and acumen, the way strings of social learning unfold is determinant to comprehend how disciplines interrelate (which may even entail hindering transdisciplinarity) as well as the scope and limitations of combining research methods from different disciplines to analyse phenomena from a spatial angle. For instance, in my proposed conceptual delimitation of social learning, I underscore that whether knowledge is produced, developed or exchanged depends on the conditions under which the interactivity underpinning the social learning unfolds (distribution of resources, power asymmetries, objectives sought,…). In view of the experience I had contributing to the production of master plans and, more specifically, examining the state of green open-space areas, an asymmetry within the power relation—exerted on the grounds of 'disciplinary canonisation'—impeded the production of new knowledge. Furthermore, by equating the spatial analysis of the green open-space areas with the mere reproduction of their geo-localisation and inventory, a whole dimension of the study containing the local knowledge and acumen of people was written off. In so doing, the social learning process was also being steered towards corroborating and thus legitimising already existing knowledge. Following von Schönfeld et al. (2020: 412), this attests to the need to relativise the somewhat taken-for-granted idea that social learning will always give way to anticipated and positive outcomes for all parties involved.

Both conceptual and practical clarity is thus crucial to pragmatise expectations and optimise the use of resources, when applying social learning as an approach in spatial and transdisciplinary research. Not only can social learning have various meanings and significances, but there is also an important distinction to be made between transdisciplinarity in research and in practice. As to the former, it may well be that, rather than transdisciplinarity, it is intra-, multi- or interdisciplinarity what is actually occurring. As expounded in the case of the *planes reguladores*, for the sake of coping with deadlines and given the resources at hand, multidisciplinary was used as an expedient at the expense of a limited interaction between disciplines and lowering the range and impact of spatial research. Regarding transdisciplinarity in practice, the valuable contribution of non-scientific actors to tie societal and research problems by combining academic and extra-academic knowledge, as proposed by Jahn et al. (2012: 8), renders tangled the basic ideas of *discipline* and *disciplinary knowledge*. Furthermore, more often than not, scientists in sustainability research projects and practical implementation initiatives (e.g. pilot projects) tend to take over the problem framing (Wuelser & Pohl, 2016). Hence the rather imperative need for non-academic participants and their knowledge to be taken seriously and fully integrated throughout the whole research or project process to, for instance, make critical decisions like the task allocation, selection of research methodology, etc. (Herrero et al., 2019: 4–5). This, moreover, begs other critical questions such as how and by which means is knowledge made valid, legitimate or practical, and how static or, conversely, susceptible to modification are the settings within which transdisciplinarity in research and/or in practice plays out.

All in all, given the pressing need to more effectively tackle the long-standing wicked problems, the effects of which have uneven impact throughout the world, it is important to realise that formulaic, overarching solutions have lapsed. To be sure, social learning does offer a lot of potential to trigger the design of policies and implementation of concrete measures that, by drawing on a *transformative transdisciplinary knowledge*, are attuned to the context wherefrom they emerge and in which they take effect. However, considering the various implications previously discussed, this should not be conceived as straight-line movement from point 'a' to point 'b'. Instead, to bring about sustainability in cities and communities through *transformative* change by dint of transdisciplinarity and spatial research and practice, coordination of efforts will have to muddle through and respond by iteratively adapting to the research specificities (from deep-seated traditions to the idiosyncrasy of researchers) and implementation practicalities (from legal and regulatory frameworks to clash of interests to lack of political will). Eventually, transdisciplinarity and spatial research, by way of social learning (and recognising its ingrained shortcomings), can result in sustainability and systemic change becoming two sides of the same coin. While prospects under the current global scenario admittedly appear dim and there is ostensibly no ample margin for experimentation, it is still worth trying out new alternatives. At the same time, we must grapple with the fact that the Earth is not a cornucopia, that the clock is ticking and that it is about time to stop talking the talk and start walking the walk.

References

Appel, J., & Kim-Appel, D. (2018). Towards a transdisciplinary view: Innovation in higher education. *International Journal of Teaching and Education, 6*(2), 61–74. https://doi.org/10.52950/TE.2018.6.2.004

Bandura, A. (1971). *Social learning theory*. General Learning Press.

Beers, P. J., Mierlo, B., & Hoes, A. C. (2016). Toward an integrative perspective on social learning in system innovation initiatives. *Ecology and Society, 21*(1), 33. https://www.jstor.org/stable/26270338

Bernstein, J. H. (2015). Transdisciplinarity: A review of its origins, development, and current issues. *Journal of Research Practice, 11*(1), 1–20.

Bieluch, K. H., Bell, K. P., Teisl, M. F., Lindenfeld, L. A., Leahy, J., & Silka, L. (2017). Transdisciplinary research partnerships in sustainability science: An examination of stakeholder participation preferences. *Sustainability Science, 12*(1), 87–104. https://doi.org/10.1007/s11625-016-0360-x

Caniglia, G., Luederitz, C., von Wirth, T., Fazey, I., Martín-López, B., Hondrila, K., König, A., von Wehrden, H., Schäpke, N. A., Laubichler, M. D., & Dang, D. J. (2021). A pluralistic and integrated approach to action-oriented knowledge for sustainability. *Nature Sustainability, 4*, 93–100. https://doi.org/10.1038/s41893-020-00616-z

Castillo Ulloa, I. (2016). From apophenia to epiphany: Making planning theory-research-practice co-constitutive. *plaNext—Next Generation Planning, 3*, 16–35.

Díaz, S., et al. (2019). Pervasive human-driven decline of life on earth points to the need for transformative change. *Science, 366*(6471). https://doi.org/10.1126/science.aax3100

Dumitru, A., Lema-Blanco, I., Kunze, I., Kemp, R., Wittmayer, J., Haxeltine, A., García-Mira, R., Zuijderwijk, L., & Cozan, S. (2017). *Social learning in social innovation initiatives: Learning about systemic relations and strategies for transformative change* (TRANSIT Brief 4). TRANSIT: EU SHH.2013.3.2-1 Grant agreement no.: 613169.

Folke, C., Colding, J., & Berkes, F. (2003). Synthesis: Building resilience and adaptive capacity in social-ecological systems. In F. Berkes, J. Colding, & C. Folke (Eds.), *Navigating social-ecological systems: Building resilience for complexity and change* (pp. 352–387). Cambridge University Press.

Folke, C., Biggs, R., Norström, A., Reyers, B., & Rockström, J. (2016). Social-ecological resilience and biosphere-based sustainability science. *Ecology and Society, 21*(3), 41. https://doi.org/10.5751/ES-08748-210341

Griggs, D., Stafford-Smith, M., Gaffney, O., Rockströn, J., Öhman, M., Shyamsundar, P., Steffen, W., Galser, G., Kanie, N., & Noble, I. (2013). Sustainable development goals for people and planet. *Nature, 495*, 305–307. https://doi.org/10.1038/495305a

Hadorn, G. H., Bradley, D., Pohl, C., Rist, S., & Wiesmann, U. (2006). Implications of transdisciplinarity for sustainability research. *Ecological Economics, 60*(1), 119–128. https://doi.org/10.1016/j.ecolecon.2005.12.002

Hadorn, H. G., Hoffmann-Riem, H., Biber-Klemm, S., Grossenbacher-Mansuy, W., Joye, D., Pohl, C., Wiesmann, U., & Zemp, E. (Eds.). (2008). *Handbook of transdisciplinary research*. Springer Science.

Hegger, D., Lamers, M., Van Zeijl-Rozema, A., & Dieperink, C. (2012). Conceptualising joint knowledge production in regional climate change adaptation projects: Success conditions and levers for action. *Environmental Science & Policy, 18*, 52–65. https://doi.org/10.1016/j.envsci.2012.01.002

Herrero, P., Dedeurwaerdere, T., & Osinki, A. (2019). Design features for social learning in transformative transdisciplinary research. *Sustainability Science, 14*, 751–769.

Holden, M., Esfahani, A. H., & Scerri, A. (2014). Facilitated and emergent social learning in sustainable urban redevelopment: Exposing a mismatch and moving towards convergence. *Urban Research and Practice, 7*(1), 1–19. https://doi.org/10.1080/17535069.2014.885735

Ingold, T. (2014). That's enough about ethnography! *HAU: Journal of Ethnographic Theory, 4*(1), 383–395.

Instituto Nacional de Vivienda y Urbanismo (INVU). (2006). *Manual de procedimientos para la redacción y elaboración de planes reguladores*. Dirección de Urbanismo.

Jahn, T. (2008). Transdisciplinarity in the practice of research. In M. Bergmann & E. Schramm (Eds.), *Tranzdisziplinäre Forschung: Integrative Forschungsprozesse verstehen und bewerten* (pp. 21–37). Campus Verlag.

Jahn, T., Bergmann, M., & Keil, F. (2012). Transdisciplinarity: Between mainstreaming and marginalization. *Ecological Economics, 79*, 1–10. https://doi.org/10.1016/j.ecolecon.2012.04.017

Jantsch, E. (1970). Inter- and transdisciplinary university: A systems approach to education and innovation. *Policy Sciences, 1*, 403–428. https://doi.org/10.1007/BF00145222

Kemp, R., Loorbach, D., & Rotmans, J. (2009). Transition management as a model for managing processes of co-evolution towards sustainable development. *International Journal of Sustainable Development & World Ecology, 14*(1), 78–91. https://doi.org/10.1080/13504500709469709

Lang, D. J., Wiek, A., Bergmann, M., Stauffacher, M., Martens, P., Moll, P., & Thomas, C. J. (2012). Transdisciplinary research in sustainability science: Practice, principles, and challenges. *Sustainability Science, 7*(1), 25–43. https://doi.org/10.1007/s11625-011-0149-x

Lawrence, R. J. (2010). Beyond disciplinary confinement to imaginative transdisciplinarity. In V. A. Brown, J. H. Harris, & J. Y. Russell (Eds.), *Tackling wicked problems: Through the transdisciplinary imagination* (pp. 16–30). earthscan.

Lawrence, M. G., Williams, S., Nanz, P., & Renn, O. (2022). Characteristics, potentials and challenges of transdisciplinary research. *One Earth, 5*(1), 44–61.

Lubchenco, J. (1998). Entering the century of the environment: A new social contract for science. *Science, 279*(5350), 491–497. https://doi.org/10.1126/science.279.5350.491

Mäntysalo, R. (2005). Approaches to participation in urban planning theories. In I. Zetti & S. Brand (Eds.), *Rehabilitation of urban areas* (pp. 23–38). University of Florence.

Markard, J., Raven, R., & Truffer, B. (2012). Sustainability transitions: An emerging field of research and its prospects. *Research Policy, 41*(6), 955–967. https://doi.org/10.1016/j.respol.2012.02.013

Nicolescu, B. (2002). *Manifesto of Transdisciplinarity*. State University of New York Press.

Nicolescu, B. (2014). Multidisciplinarity, interdisciplinarity, indisciplinarity, and transdisciplinarity: Similarities and differences. *RCC Perspectives, 2*, 19–29.

Nowotny, H., Scott, P., Gibbons, M., & Scott, P. B. (2001). *Re-thinking science: Knowledge and the public in an age of uncertainty*. Polity Press.

Pahl-Wostl, C. (2006). The importance of social learning in restoring the multifunctionality of rivers and floodplains. *Ecology and Society, 11*(1), 1–10. https://www.jstor.org/stable/26267781

Parson, E. A., & Clark, W. C. (1995). Sustainable development as social learning: Theoretical perspectives and practical challenges for the design of a research program. In L. H. Gunderson, C. S. Holling, & S. S. Light (Eds.), *Barriers and bridges to the renewal of ecosystems and institutions* (pp. 428–460). Columbia University Press.

Plummer, R., Blythe, J., Gurney, G. G., Witkowski, S., & Armitage, D. (2022). Transdisciplinary partnerships for sustainability: An evaluation guide. *Sustainability Science, 17*, 955–967. https://doi.org/10.1007/s11625-021-01074-y

Popa, F., Guillermin, M., & Dedeurwaerdere, T. (2015). A pragmatist approach to transdisciplinarity in sustainability research: From complex systems theory to reflexive science. *Futures, 65*, 45–56. https://doi.org/10.1016/j.futures.2014.02.002

Popa, S., Soto-Acosta, P., & Martinez-Conesa, I. (2017). Antecedents, moderators, and outcomes of innovation climate and open innovation: An empirical study in SMEs. *Technologial Forecasting and Social Change, 118*, 134–142. https://doi.org/10.1016/j.techfore.2017.02.014

Ramadier, T. (2004). Transdisciplinarity and its challenges: The case of urban studies. *Futures, 36*(4), 423–439.

Ravetz, J. R. (1999). What is post-normal science. *Futures, 31*, 647–653.

Reed, M. S., Evely, A. C., Cundill, G., Fazey, I., Glass, J., Laing, A., Newig, J., Parrish, B., Prell, C., Raymond,

C., & Stringer, L. C. (2010). What is social learning? *Ecology and Society, 15*(4). https://doi.org/10.5751/es-03564-1504r01

Rittel, H. W., & Webber, M. M. (1973). Dilemmas in a general theory of planning. *Policy Sciences, 4*, 155–169. https://doi.org/10.1007/BF01405730

Rydin, Y. (2010). *Governing for sustainable urban development*. Earthscan.

Scholz, G., Dewulf, A., & Pahl-Wostl, C. (2014). An analytical framework of social learning facilitated by participatory methods. *Systemic Practice and Action Research, 27*(6), 575–591. https://doi.org/10.1007/s11213-013-9310-z

Shove, E., Pantzar, M., & Watson, M. (2012). *The dynamics of social practice: Everyday life and how it changes*. Sage.

Slater, K., & Robinson, J. (2020). Social learning and transdisciplinary co-production: A social practice approach. *Sustainability, 12*(18), 1–16. https://doi.org/10.3390/su12187511

Sol, J., Van Der Wal, M. M., Beers, P. J., & Wals, A. (2018). Reframing the future: The role of reflexivity in governance networks in sustainability transitions. *Environmental Education Research, 24*(9), 1383–1405. https://doi.org/10.1080/13504622.2017.1402171

Tewdwr-Jones, M. (2002). Personal dynamics, distinctive frames and communicative planning. In P. Allmendiger & M. Tewdwr-Jones (Eds.), *Planning futures: New directions for planning theory* (pp. 65–92). Routledge.

von Schönfeld, K. C., Tan, W., Wiekens, C., & Janssen-Jansen, L. (2020). Unpacking social learning in planning: Who learns what from whom? *Urban Research and Practice, 13*(4), 411–433.

Wells, J. (2012). *Complexity and sustainability*. Routledge.

Wickson, F., Carew, A.L. and Russell, A.W. (2006). Transdisciplinary research: Characteristics, quandaries and quality. *Futures 38*(9), 1046–1059. https://doi.org/10.1016/j.futures.2006.02.011

Wuelser, G., & Pohl, C. (2016). How researchers frame scientific contributions to sustainable development: A typology based on grounded theory. *Sustainability Science, 11*(5), 789–800. https://doi.org/10.1007/s11625-016-0363-7

Young, H. P. (2009). Innovation diffusion in heterogeneous populations: Contagion, social influence, and social learning. *The American Economic Review, 99*(5), 1899–1924. https://doi.org/10.1257/aer

Ignacio Castillo Ulloa, PhD, is lecturer and researcher at the Chair of Urban Development and Urban Design, Department of Urban and Regional Planning, and scientific coordinator of the Global Center of Spatial Methods for Urban Sustainability (SMUS), Technische Universität Berlin. He is co-editor of the book Spatial Transformations published by Routledge. He is also a research associate at the Collaborative Research Centre 1265 'Re-Figuration of Spaces'. His research interests include socio-spatial development and alternative disruptive (local) practices, critical urban research, transdisciplinarity and the use of spatial research methods.

Open Access This chapter is licensed under the terms of the Creative Commons Attribution 4.0 International License (http://creativecommons.org/licenses/by/4.0/), which permits use, sharing, adaptation, distribution and reproduction in any medium or format, as long as you give appropriate credit to the original author(s) and the source, provide a link to the Creative Commons license and indicate if changes were made.

The images or other third party material in this chapter are included in the chapter's Creative Commons license, unless indicated otherwise in a credit line to the material. If material is not included in the chapter's Creative Commons license and your intended use is not permitted by statutory regulation or exceeds the permitted use, you will need to obtain permission directly from the copyright holder.

3

Participation in Transdisciplinary Urban Planning Practice and Research: Spatial Methods in Action

Inês Martina Lersch, Angela Million, and Jacqueline Custódio

3.1 Introduction: Linking the World of Participation and Transdisciplinarity for SDG11

Participation and collaboration in urban planning have witnessed a significant shift and intensification in the Global North since the late 1990s (Forester, 1999; Healey, 1993). By the same token, these principles have taken root in an adapted manner in the Global South (Watson, 2014), emerging as crucial responses to the challenges posed by urban development. Moreover, the involvement of stakeholders and the public is paramount in shaping cities and communities that are inclusive, safe, resilient and sustainable.

Such engagement ensures that a diverse array of perspectives, needs and concerns are thoroughly considered throughout the planning and implementation processes. Therefore, the participatory and collaborative paradigm could also be seen as a foundational element for advancing Sustainable Development Goal 11 (SDG11).

This chapter delves into SDG11 by considering that it lies at the heart of urban planning education and research. SDG11 is a fundamental component in many existing curricula and scholarly discussions within urban planning and, therefore, an integral goal of urban planning education and practice. Drawing upon successful experiences derived from the practical implementation of SDG, it becomes evident that participation plays a crucial role in fostering ownership, accountability and legitimacy in the pursuit of the goals outlined within SDG11 (Smit et al., 2019; Croese et al., 2021; Barnerjee, 2021). Simultaneously, in tackling the intricate challenges of urban development, we deem it crucial to transcend disciplinary boundaries and promote collaboration among diverse stakeholders, both in practice and research.

Recognising this imperative, we must not only embrace the concepts of transdisciplinarity and participation but also bridge the gap between two worlds that do not always readily converge. Through the integration of transdisciplinary and participatory approaches, we can formulate strategies for achieving SDG11 that are more informed, comprehensive and people-centred. This synergy not only has the potential to bridge the gap between research and practice but also to ensure that urban development initiatives are

---

I. M. Lersch
Department of Urban Planning + SMUS, Federal University of Rio Grande do Sul, Porto Alegre, Brazil
e-mail: martina.lersch@ufrgs.br

A. Million (✉)
Department of Urban and Regional Planning + SMUS, Technische Universität Berlin, Berlin, Germany
e-mail: million@tu-berlin.de

J. Custódio
Department of Urban Planning, Federal University of Rio Grande do Sul, Porto Alegre, Brazil

© The Author(s) 2025
F. Frehse et al. (eds.), *Spatial Methods in Transdisciplinarity for Urban Sustainability*, Sustainable Development Goals Series, https://doi.org/10.1007/978-3-031-84367-9_3

both sustainable and responsive to the aspirations and needs of the communities they serve. Collaborative efforts can additionally lead to innovative solutions that take the diverse perspectives of stakeholders into account.

Looking at the role of urban planning and urban planners, the participatory and collaborative shift can be considered a valuable asset. We contend that this shift has nurtured the development of spatial methods and a broader range of formats and outcomes in transdisciplinary and participatory planning processes. Their pre-existing integration into the university education of future planners and designers serves as a pivotal element in the transdisciplinary pursuit of SDG11.

Consequently, the objectives of this chapter are threefold: first, to explore the role of participation in transdisciplinary urban planning research and practice; second, to examine urban planning pedagogy and the spatial methods applied in this context; and third, to discuss the challenges and potential shortcomings associated with participation in transdisciplinary urban research and practice by drawing on insight from an exemplary case study: the preservation and so-called revitalisation of the old port area of the Southern Brazilian city of Porto Alegre.

This chapter is based on our individual approaches to teaching, researching and practicing urban planning and urban design within the domain of higher education and planning practice. With over two decades of experience in planning and research, Angela Million has been actively engaged in academic scholarship at Technische Universität Berlin (TU Berlin). She co-founded an NGO dedicated to built environment education and the participation of young people in urban planning and architecture. Her research focuses on the outcomes and impact of participatory urban planning and cities as educational settings. Inês Martina Lersch contributes with her teaching and research expertise in urban heritage and social housing at the Federal University of Rio Grande do Sul (UFRGS) alongside a tradition of academic involvement in local and national urban policy and planning practice. She has additionally served as a guest scholar at TU Berlin. Both share a teaching tradition that involves training future urban planners and architects through university outreach and live projects in communities, towns and cities. Jacqueline Custódio, in turn, is an accomplished lawyer with more than a decade of experience in the field of cultural heritage and holds a master's degree in museology and heritage. Custódio, as a practitioner, alongside Lersch as an academic, actively participated in urban development processes and public discussions, notably regarding the preservation and revitalisation of the historic downtown area of Porto Alegre, particularly the old port area, from 2021 to 2023.

3.2 Participation in Transdisciplinary Research and Practice

When delving into the recent scholarly literature, the terms participation, transdisciplinary and even co-production are often used interchangeably, at times as synonyms (Schäpke, 2017; Fritz & Binder, 2018), often creating challenges in distinguishing between these concepts and discussions. However, there is a rich body of literature specifically devoted to transdisciplinary research and an even larger one on participatory research. Participatory Research (PR) or Participatory Action Research (PAR) is a research methodology wherein researchers and participants collaborate to study issues and produce knowledge to bring change (Reason & Bradbury, 2008; Argyris & Schön, 1989). With a longstanding history in Latin America (Fals-Borda, 1987; Streck & Rodrigues Brandão, 2005), PR has evolved into an umbrella term in academia, encompassing all research approaches that engage non-academic stakeholders in each step of the research process, aiming to apply research results to problem-solving. As such, PR explicitly mandates transdisciplinary collaboration. While it does not always involve collaboration between researchers from different disciplines, it does include community members and stakeholders with expertise in different areas. Consequently, PR is often referred to and cited as a transdisciplinary research approach.

Concerning transdisciplinary research, Pohl (2010) analysed its characteristics and identified four key features, which are often cited by scholars: (a) Relating to socially relevant issues, (b) Transcending and integrating disciplinary paradigms, (c) Participatory research and d) Searching for a unity of knowledge. Pohl emphasises that these features carry different weight in various forms of transdisciplinary research and defines three concepts of transdisciplinary research as combinations of these four characteristics (a, b, c, d). Only in one of Pohl's concepts, concept B, does the participation of non-academic actors, such as stakeholders from science, civil society and the private and the public sector have an active role: 'In the early years of the new millennium, concept B – and specifically the feature of participatory research – gained still more momentum. In some of the research programs, transdisciplinarity even became synonymous with participatory research' (Pohl, 2010, 68). This shows that participation as a characteristic of transdisciplinary research cannot always be expected to be found. However, its significance is on the rise compared to the early years of transdisciplinary research.

When examining the process of an ideal-typical transdisciplinary research process, as discussed by Bergmann, 2010; Lang et al., 2012, several parallels with participatory research emerge. Initially, the problem framing and research design are collaboratively defined among stakeholders. Subsequently, a process of co-production involving the development of application-oriented solutions and knowledge ensues, with the ultimate goal of applying the generated knowledge or implementing it in strategies and actions. Similar to PR, the degree of stakeholder involvement can vary in transdisciplinary research projects. This aspect has received considerable attention in the research and reflection on transdisciplinary processes (Brandt et al., 2013; Schäpke, 2017; Zscheischler et al., 2014; Mobjörk, 2010). Here, the dynamics of different participation intensities are described ranging from information via consultation, cooperation and collaboration to empowerment within one transdisciplinary research project (Stauffacher et al., 2008). This analytical lens will also guide our examination of the participatory planning process of the Porto Alegre harbour front.

While participation in transdisciplinary research has been acknowledged and investigated, it is worth noting that intensive forms of involving non-scientific actors, including empowerment, are relatively rare in the practice of transdisciplinary research, a point criticised by other scholars. When looking at the reasons for this, participatory transdisciplinary research processes (Mobjörk, 2010; Lang et al., 2012; Scholz & Steiner, 2015) face similar challenges as participatory research (Reason & Bradbury, 2008; Bergold & Thomas, 2012), which include:

- *Complexity*: research projects incorporating multiple disciplines and perspectives can become intricate and difficult to manage. It can be challenging to integrate diverse perspectives and knowledge systems and to find common ground between different disciplinary approaches.
- *Communication*: communication can be difficult when researchers and stakeholders come from different disciplinary backgrounds. The use of technical jargon and specialised language can create barriers, impeding progress on the project.
- *Power dynamics*: power dynamics can be a significant challenge in participatory transdisciplinary research. Disciplinary hierarchies and power imbalances can create tension and hinder effective collaboration between researchers and stakeholders.
- *Resource constraints*: research projects that involve multiple disciplines and stakeholders may require additional resources in terms of time, funding and personnel. Limited availability of these resources can constrain the scope and impact of the project.
- *Output, outcomes and their evaluation*: evaluating the success of a participatory transdisciplinary project is a complex task, given that success may be defined differently across different disciplines and perspectives.

Despite these challenges, in their comparative analysis on effects of 101 transdisciplinary research projects in the Global South and North Pärli, Fischer and Lieberherr 2022 came to the conclusion that those projects have the potential to not only generate knowledge but also solutions and outputs to complex problems of sustainable development and to promote more inclusive and equitable research practices and more trust among participants. Given variations in historical evolution and circumstances, their findings reveal that North–South projects displayed a greater impact on 'societal effects and uptake of knowledge' (Pärli, Fischer, and Lieberherr 2022), whereas projects in the Global North yielded more traditional academic outcomes, such as scholarly publications. In terms of the nexus of outcomes, North–South initiatives placed a heightened emphasis on inclusivity and participation compared to those in the Global North. The findings of Pärli et al. can also be interpreted as a necessity, wherein the outputs of transdisciplinary processes need to be redefined, expanding beyond academic papers and encompassing a wider variety, especially in light of increased participation.

3.3 Participation in Urban Planning Pedagogy and the Role of Spatial Methods

As previously mentioned, public participation and collaboration in planning have garnered increasing attention over the last few decades. This shift from positivistic and technocratic approaches in planning theory to post-positivistic standpoints has led to the emergence of various 'turns', such as 'communicative planning' and 'argumentative planning' (Healey, 1993; Forester, 1999; Fischer & Forester, 1993), as well as 'collaborative planning' (Healey, 1997). These approaches and perspectives represent the major frameworks for understanding participation in urban planning theory and practice: The 'communicative turn' in the 1990s marked a (debatable) shift in urban design and urban planning (Yiftachel & Huxley, 2000). This shift transformed the perception of these fields from mere 'Gestaltungsakt' (design acts) to processes that involve negotiating diverse needs and preferences of various stakeholders. Questioning this Global North perspective, Watson (2016; 2014) underscores the substantial disparities between planning practices in the Global North and the Global South. She stresses that much of the analytical and normative thinking on state-society engagement in planning has been implicitly informed by the particular social conditions of Western liberal democracies and mature advanced economies. Watson (2014) further asserts that participation in the Global South often occurs 'under conditions in which governments are either unwilling or unable to deliver land and services' (Watson, 2014, 63). Despite differences, Watson also identifies commonalities. She sees worldwide shared scholarly concerns regarding sustainable urban development. Watson notes, 'all these positions have been concerned with how state and society can engage in order to improve the quality of life of populations, sometimes with an emphasis on the poor and marginalised, and sometimes with these outcomes specified as socio-spatial justice and more equitable and sustainable outcomes of state intervention in urban development, and how professionals can act to promote this. Certainly, underlying goals and values have much in common' (Watson, 2014, 69). Regarding the differences, Watson highlights a prevalence of planning processes in the Global South that, in comparison to the participatory planning processes she analysed in the context of participation in slum housing development, tend to operate outside of formal channels, filling gaps in formal planning, with a greater emphasis on implementation and management, where power and conflict become central issues. Additionally, there is an emphasis on less talking and more direct action and implementation.

In all of these planning processes, built environment professionals and planners often transition from their traditional role as technical experts to that of providing community support. As our case will demonstrate later in this chapter, they frequently collaborate with colleagues from the cultural, social or legal sectors. These professionals may work within NGOs, city administrations or academic institutions. Urban planning and urban studies academics, such as ourselves, often hold various roles simultaneously, serving as university scholars, educators, practitioners and activists. Professionals may also take evolving roles within participatory planning and research processes, including teaching responsibilities such as training communities to interpret maps and satellite imagery and understand financial aspects. It is common for university students in urban planning and architecture programmes to participate in such processes, often through outreach and live projects (Watson, 2014, 69). These projects, conducted in collaboration with communities or as joint studio projects between NGOs and planning students, are prevalent in urban planning and architecture higher education. Similar academic engagements that combine research and practice within other disciplines exist in various forms in higher education worldwide (Berry & Chisholm, 1999). They are often referred to as university extension or outreach projects, while their pedagogy is associated with service-learning. In planning and architecture university programmes, they may be known as life projects (Sara, 2011; Sara & Jones, 2018) or joint (studio) projects (Watson & Odendaal, 2013) and practicum (Kotval, 2003). Latin America has a longstanding and robust tradition of project practices of this nature. In Brazil, for example, they are called *Projetos de Extensão*. Scholars such as D'Ottaviano, Rovati, 2017, 2019 and Mello et al. 2019 have primarily applied them in the field of urban and regional planning education. These experiences equip students with spatial methods knowledge and prepare them for the collaborative nature of the planning discipline.

As discussed earlier, contemporary spatial planning in many communities strives towards a collaborative nature and the incorporation of participatory processes. To this end, planners employ a spectrum of methods throughout their urban design and planning process. From our perspective, all the methods falling within these fields of application are inherently spatial. They are used to investigate spatial questions and engage with the public concerning spatial issues and conceptualising solutions. We propose categorising these spatial methods based on their application process, specifically as spatial methods used for (i) analysis, (ii) conceptualising and (iii) stakeholder discussion and interaction based on (Barton, 2010; Curdes, 1995; Giseke, 2021; Gehl, 1987; Heinrich et al., 2021).

(i) Analysis methods play a pivotal role in understanding the context, character and dynamics of urban environments across various scales. They involve identifying pertinent data, space-related qualities, interests and needs to establish a foundation for conceptualising urban actions or plans. Spatial research covers a broad spectrum, including history, physical form, accessibility, social relations and activities. Traditional spatial methods such as statistical surveys, spatial mapping, cartography, interviews, photography and video, along with 2D and 3D models, sketches, infographics, mathematical models and other analogue and digital quantitative tools, are essential.

(ii) The conceptualisation process—often referred to as the urban planning or urban design process—itself, and if separable from the analytical phase, relies on evidence, analysis, empirical investigation and intuition (Lawson, 2009). The evaluation of ideas and assessment of solutions can be also guided by analytical methods embedded within the contextual, theoretical framework, drawing from a range of diverse disciplines such as ecology and political studies (e.g. political ecology), sociology (e.g. actor-network theory), mathematics and computer science (cybernetics). Integral to the conceptualisation process are representation and design simulation tools that facilitate the articulation of visions, frameworks and policies. They lay out and often visualise comprehensive design strategies that describe, coordinate and apply quality design intentions in complex urban situations. Integrated urban development strategies leverage a combination of these tools to implement urban projects across various phases. The tools produce outcomes and encompass sketches, infographics, maps, drawings (including master plans), action plans, work-in-progress models, computer-aided design (CAD) plans, images, collages, photography, audio, film, comics, choreography, narratives and games, both in analogue and digital formats, to name a few.

(iii) Lastly, discussion and communication among stakeholders facilitate interaction, dialogue and the integration of feedback into further conceptualising iterations. Spatial methods encompass a range of formats customised for stakeholder participation and specific issues. These formats include moderated discussions, charrettes, workshops, artistic performances and games. The goal is to facilitate collaborative decision-making in the design process. Media and modalities that enable live interaction and communication and contribute to the development and identification of perspectives regarding problems, tasks and solutions.

It is important to note that these three groups of spatial methods, as well as the described fields of action, are not strictly isolated. Analysis, design, planning steps and stakeholder communication are interdependent, interconnected and often overlap. Furthermore, these phases are often iterative, and the process of conceptualisation itself can be a communicative action. This is also why the editors of the urban design method book (Giseke, 2021) did not organise spatial methods into a chapter structure, as most methods cannot be strictly categorised into a single field of action. Also, the list of spatial methods and formats mentioned here is by no means exhaustive. Spatial methods are frequently combined, adapted and contextualised (e.g., Million, 2021). Additionally, the spatial methods applied in diverse formats result in various outcomes that extend well beyond maps, reports and academic papers. It also can include temporary installations, blog posts, songs and theatrical performances.

Examining scholarship on methodological and methodical research approaches reveals a historical trend: urban planners, designers and architects have had limited participation in debates and publications, particularly when compared to other disciplines, especially those within the social sciences (Heinrich et al., 2021, 9–10). However, it is worth noting that they possess a comprehensive set of methods for spatial research and have recently started to engage in methodological discourse (ibid.). This situation has been spurred by recent publications on spatial methodology (Giseke, 2021; Heinrich et al., 2021), in which Angela Million has also participated, which brings interdisciplinary expertise together to provide an overview of qualitative and visual methods in spatial research. Both editorial teams of the two forementioned spatial methodology books highlight the evolving relationship between society and science, emphasising the need for new forms of knowledge production to address spatial challenges. In this context, they underscore the significance of participatory approaches using spatial methods in contemporary times. This aspect will be further emphasised through our case study of Porto Alegre.

3.4 Spatial Methods in Participatory Transdisciplinary Research: Historic Waterfront Development in Porto Alegre, Brazil

The project and proposal known as the '*Cais Cultural: Proposal to occupy the Porto Alegre pier*' is the case study analysed in this chapter. The city of Porto Alegre gained international recognition in the academic field of public management, which also encompasses urban planning, in the 1990s and early 2000s for its participatory budgeting programme (Weyh & Streck, 2003). This initiative empowered residents to participate directly in the allocation of municipal resources. However, recent administrations have adopted practices that run counter to the principles of popular participation. In the case presented here, we delve into the current transdisciplinary and participatory complexities surrounding the preservation and revitalisation of Porto Alegre's historic urban centre, with a specific focus on the old port area. In the context of SDG11, this case study sheds light on the aspirations, challenges and limitations of participation and transdisciplinarity within urban development, emphasising the city's cultural heritage and promoting an environment characterised by respect and inclusivity. Spatial methods occupy a central position in this discourse, serving as connectors between two worlds and offering opportunities for their combined use in diverse formats.

3.4.1 Porto Alegre's Cais do Porto in the Context of SDG11

Founded in 1772 by Azorean immigrants along the shores of the Guaíba Lake, Porto Alegre initially served military and port functions under Portuguese rule. Presently, as the capital of the southern state of Rio Grande do Sul (RS), Porto Alegre has 1.330 million inhabitants and a demographic density of 634.5 inhabitants per square km (IBGE, 2022). Regarded as the foundational cornerstone of the city and holding immense historical significance due to its lasting impact on the economic dynamics of Porto Alegre (Souza & Müller, 2007; Alves, 2005), the *Cais do Porto* (harbour) has been at the centre of disputes over the past three decades. On one side are those advocating for its privatisation, while on the other are those pushing for its public and democratic occupation. Built in 1913, expanded in 1927, and deactivated in 1994 for shipping activities, the object of interest and dispute corresponds to a built area of 4.4 ha with 11 warehouses and free space of 12.1 ha, amounting to 16.5 ha, near the city's historic centre.

As previously mentioned, Porto Alegre was once globally recognised as a city where public authorities actively engaged in open dialogue with society, evaluating priorities and addressing the needs of various urban regions. Those practices, however, changed over time. In 2020, the state government hired the National Development Bank (BNDES) and private consultancies to decide how to transform the port quay, considered a cultural, material and immaterial heritage of Porto Alegre, into a commercial area and real estate venture. The proposal sparked a social movement asserting the right to participate in decisions regarding the site that represents the city's 'identity card', as its visually striking structure uniquely identifies Porto Alegre for those arriving by plane or boat (Fig. 3.1).

In light of the above facts, we find a clear nexus between our present research and the field of law. Citizenship in Brazil was only officially ensured with the promulgation of the Federal Constitution in 1988, further solidified by the enactment of a specific law called '*Estatuto da Cidade*' (City Statute) in 2001. In the recent era of Brazilian democracy, established by the Federal Constitution of 1988, citizenship stands as a fundamental principle. In this context, citizenship is defined as the active participation of each individual in the political life of the state, ensuring that everyone possesses a voice in a democratic society. This seemingly straightforward aspect was a monumental achievement for the Brazilian people, particularly in the aftermath of two decades of military dictatorship. The foundational document, the Democratic Rule of Law, underscores citizenship as one of its pillars and the condition for active popular participation in public life and recognises the significance of civil society's demands. In turn, the second document mentions popular participation, particularly within the realm of urban planning.

The 'Cais Cultural: Proposal to occupy Porto Alegre's pier' project was a laborious, at times conflicting, instigating and highly invigorating process between the University and the Collective '*Cais Cultural Já*'. This initiative was forged through collective participation, interdisciplinarity between areas of academic-scientific knowledge, and transdisciplinarity by incorporating knowledge from different social groups. The inception of this project emanated from a collective of non-academic stakeholders, comprising artists and professionals from various cultural

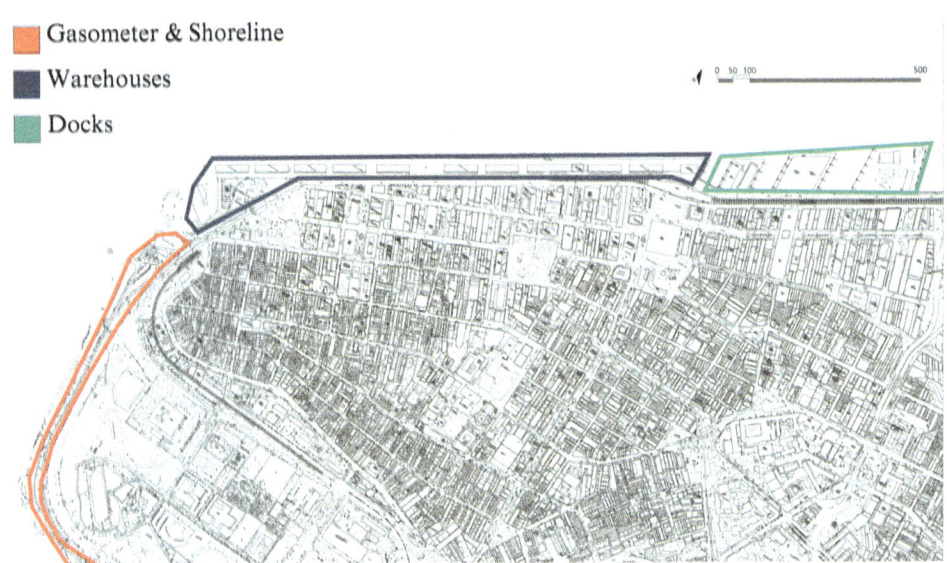

Fig. 3.1 Map of the pier area. Source: Shapefiles Prefeitura Municipal de Porto Alegre processed in SIG Qgis

sectors: scenic (theatre, dance, music), literature, visual arts, audio-visual, handcrafts, as well as popular cultures originating from Indigenous peoples, African populations and the immigrant community.

On the academic front, this study is the result of research conducted by professors, graduate students and undergraduate students collaboratively engaged in outreach projects, in a dialogical relationship between the University, Society and a research group of the Federal University of Rio Grande do Sul (UFRGS). Owing to the constraints imposed by the COVID-19 pandemic, a substantial portion of the work until November 2021 was carried out remotely. At times, in-person meetings, including workshops, meetings and hearings, were facilitated when necessary and feasible.

The problem under consideration was: how to guarantee the public and democratic use of Porto Alegre's Pier? This issue exposes the deviation from the urban planning principles advocated by the UN, which encourages the transformation of cities into sustainable urban centres. In the case study, the government's project proved to be privatising, gentrifying and elitist, with detrimental effects on the environment, cultural heritage and landscape that characterises Porto Alegre. Hence, the formulation of guidelines for the occupation of the waterfront emerged as a proactive measure by the cultural community to propose alternatives for space production, in alignment with the third goal of SDG11, which focuses on promoting inclusive and sustainable urbanisation (Fig. 3.2).

The proposal, designed to ensure the public use of the site, hereby understood as universal and free access for the entire population to Cais de Porto Alegre, enabled a public debate about the project involving the citizens of Rio Grande do Sul. The proposal (Marzulo et al., 2021) presented feasibility studies on the use and occupation of Porto Alegre's pier for cultural activities, as well as guidelines for the viability and financial sustainability of the enterprise and management through participatory governance mechanisms. Proponents understood that the proposal would undergo an extensive discussion process with instances involving citizen participation, social and community organisations, and professional and technical entities, as well as governmental and non-governmental institutions involved in the issue. The proposal sought to adhere to the principles of the right to the city and life, as advocated by the City Statute, the federal legislation on urban policy and the Federal Constitution of Brazil.

Fig. 3.2 The pier and the press conference, one of the moments in which the proposal with guidelines was presented and discussed in November 2021 (© Lersch, Custódio, 2021)

The exercise of thought that generates intelligence presupposes education, culture, science and art. Believing in this principle, we developed the present proposal, the fruit of joint elaboration, collective work, and much dialogue between teachers, researchers, students, producers of culture, artists and the community. Times are hard, and that is why collective construction and articulation between scientific knowledge and popular and traditional knowledge (…) are even more necessary so that we may build a more dignified and better life for all. Furthermore, constructing a dignified planetary life starts locally, in our village, our city. (Marzulo et al., 2021)

Returning to the recommendations of SDG11, target 11.3 outlines three pillars with concrete recommendations to overcome challenges and achieve objectives. This research-practice project aligns with two of these pillars: a) 'inclusive governance; inclusive urban planning; citizenship; fostering participation, transparency, and democratization', as well as b) 'recognition of social actors – including gender – for migration and refugees; encompassing identity, cultural practice, diversity and heritage; safer cities; livelihoods, well-being; risk of poverty and employment vulnerabilities; inclusive and solidarity economy' (UN, 2015).

Regarding the development of cultural activities and heritage, target 11.4 states that 'the promotion of culturally sensitive urban strategies is essential for building resilient and inclusive cities', and that 'access to culture and participation in cultural life must be integral parts of all urban policies' (UN, 2015). Given the waterfront's location in the city's historical centre, another strategy from the noted target is worth mentioning: 'Historical centres provide living laboratories of dense urban areas with mixed functions and quality public spaces, where innovative urban approaches are tested (including soft mobility or mixed ownership) to combine conservation requirements with improved quality of life' (UN, 2015).

The reintegration of *Cais do Porto* to the city and citizenship is based on four principles to ensure that it becomes a cultural, economically diverse and creative space: (i) cultural economic diversity and creativity; (ii) environmental sustainability; (iii) sustainable management and participative governance; and (iv) integration to the historical centre.

Fundamentally, qualified public spaces, with permanent maintenance and security, attract people, generate resources, foster complementary and compatible economic activities intertwined with cultural, recreational and leisure activities, increase the valuation of the surroundings, reduce insecurity—given that higher levels of activity often correlate with reduced rates of violence, improve the quality of life in all senses and give rise to urban landscapes teeming with life. All over the world, the preservation of urban landscapes is paramount for both attracting people and affirming a distinctive image that extends beyond the immediate location, with an equally significant impact on the daily lives of the inhabitants.

3.4.2 Analysis of Spatial Methods in a Dynamic Participatory and Transdisciplinary Process

The present case study delves into the dynamics of participation within a transdisciplinary process, examining expert drivers and legal framings and providing insight into prospective policies and practices in urban development. Situated in the historical city centre, the case deals with a profoundly symbolic location for the city, one that has seen numerous proposed urban development projects for the waterfront following the cessation of its port functions. However, none of these proposals have never come to fruition, sharing the common trait of being developed as business plans without any social involvement in their conception. In the proposal preceding the current one, social movements mobilised to prevent the creation of a shopping mall and buildings along the waterfront. When the current urban planning project began to take shape, civil society adopted a proactive stance, seeking technical assistance from academia. This initiative culminated in a participatory research experience that integrated knowledge from urban planning, administration and sociology.

To discuss the participatory aspects of the transdisciplinary process, the dynamics of different intensities of participation as well as the use of methods, we employed the model proposed by Stauffacher et al. (2008). We provided an overview of the steps and participating partners in the transdisciplinary case study Porto Alegre, Brazil (Table 3.1) and analysed participation levels such as information, consultation, cooperation, collaboration and empowerment in relation to project progress and methods applied (Fig. 3.3).

Proposing to work with participatory research entails moving beyond theoretical discussions and necessitates navigating the tensions between the various stakeholders involved in the process. The following diagram illustrates the role of each actor.

The university team proposed some experiences at the pier: we (the UFRGS team) visited the pier area with representatives of the collective, we walked along the pier area and we visited the warehouses and inspected their maintenance conditions. Our observation revealed an advanced level of degradation in the warehouses, leading us to anticipate a considerable cost for revitalisation. Our efforts also involved applying methods for recognition and mapping, identifying open and accessible spaces suitable for events such as street theatre performances. As the risks of the pandemic subsided, the research team invited representatives of the collective to the university. Together with approximately 60 artists and cultural workers, we also organised an artistic parade and a cultural meeting at the pier, as a demonstration of the importance of occupying public spaces with art and culture (Fig. 3.4). Following these activities, the subsequent phase involved public hearings for negotiations with government officials and legislators (Fig. 3.5).

In Table 3.2, we describe the steps.

Adopting the model proposed by Stauffacher et al. (2008), we may express the process through the following phases, steps and methods (Fig. 3.6).

The intersection between research and practice in the presented case resulted in an initial product: a document outlining guidelines for the cultural use of the harbour warehouses. The guidelines were organised into the following parts: (a) theoretical and conceptual discussion on the social function of the cultural heritage of cities; (b) historical overview of the various interventions in the quay over the last 30 years; (c) proposal of general guidelines for the conservation of the area: the return of the quay to the city and citizenship; (d) feasibility study on the use and occupation of the quay's spaces and, finally; (e) design for the financing and management of the *Cais do Porto Cultural* project. This document, presented in the form of a booklet, provided context for the proposal, guiding principles and envisioned objectives. The booklet was distributed widely to various sectors of the society and was even given to the Minister of Culture.

Table 3.1 Actors and their roles, case study Porto Alegre, Brazil

Artists and workers of the culture field (non-academic stakeholders)	
We identified 34 groups from different cultural backgrounds, including performing arts, visual arts, audio-visual, dance, theatre, street theatre, handicrafts, music, literature and flea markets; as well as groups that work in the fields of heritage, human rights, solidarity economy and gastronomy. We also interviewed representatives from the arts, education, Afro-Brazilian culture groups and Indigenous groups. These non-academic stakeholders played a pivotal role in describing their respective activities and needs, elucidating how they envisioned utilising the spaces earmarked for occupation on the pier. Throughout the process, these stakeholders actively participated in every step, except for the writing of the document and the drawing phases (that included more technical aspects); sometimes the whole group was present, and other times, just its representatives. Two group members (we may refer to them as practitioners) were responsible for bridging the gap between the stakeholders and the research team. The collaborative process of constructing the proposal was consistently guided by the joint efforts of civil society and academia.	
University team	
The group from UFRGS was initially comprised of 4 senior researchers. The senior researchers exhibited a high level of participation throughout all phases, primarily taking on a coordinating role in the process. Additionally, their responsibilities included dedicating time to listen to the stakeholders, propose methods and articulate dialogue among the various actors. A team of students, comprising 4 graduate and 4 undergraduate students, subsequently joined the project. Their involvement included contributions to mapping, documental research and literature review phases. The students helped to conduct interviews and surveys, along with organising data and creating content for social networks on specific occasions.	
Politicians	
A state legislator and a city councillor joined the project to seek parliamentary resources and were responsible for articulating and proposing public debates. Their role was essential for creating political space and reinforcing the theme of the cultural proposal. Their level of participation was moderate, primarily focused on negotiation and advocacy efforts.	
State government	
The state secretaries played a significant role, considering that the pier area falls under the jurisdiction of the Rio Grande do Sul government. They operated as listeners and receivers of the proposal and were also responsible for promoting the democratic discussion. They had a fair level of participation, limited to the negotiation table and public hearings, where they collaborated with two managers and one architect.	
Consortium	
While not formally considered part of the research, this entity operated as another interested party in the utilisation of the pier, as it was responsible for a privatisation project. However, engagement with this entity was necessary for communication purposes. Their level of participation was very low, or essentially non-existent, as they only appeared at a few negotiation tables and public hearings, collaborating with two managers and two architects.	

3.4.3 Discussion of the Case Study, Porto Alegre, Brazil

The process of formulating guidelines for the cultural occupation of Porto Alegre's waterfront serves as a reference point for identifying both shortcomings and successes in the experience, offering valuable insight for refining the management of future similar cases. Positive achievements include the development of a concrete and feasible proposal, considering the state government's commitment to the public interest in utilising a site of great symbolic importance that had remained inaccessible and closed to the public for over a decade. From the perspective of the practitioners and social movements, the utilisation of spatial methods brought the community and the university closer together, enabling the exchange of experience, each with their expertise. The in-depth knowledge of the cause by civil society offered insight into strategies that proved effective for the group's objectives, as they had already been actively working to prevent the privatisation and gentrification of the waterfront. This input served as a basis for formulating occupation guidelines and possibilities for shared management based on academic foundations developed within the scope of university outreach projects.

The versatility of spatial methods becomes evident insofar as they transcend traditional disciplinary boundaries, finding applicability in

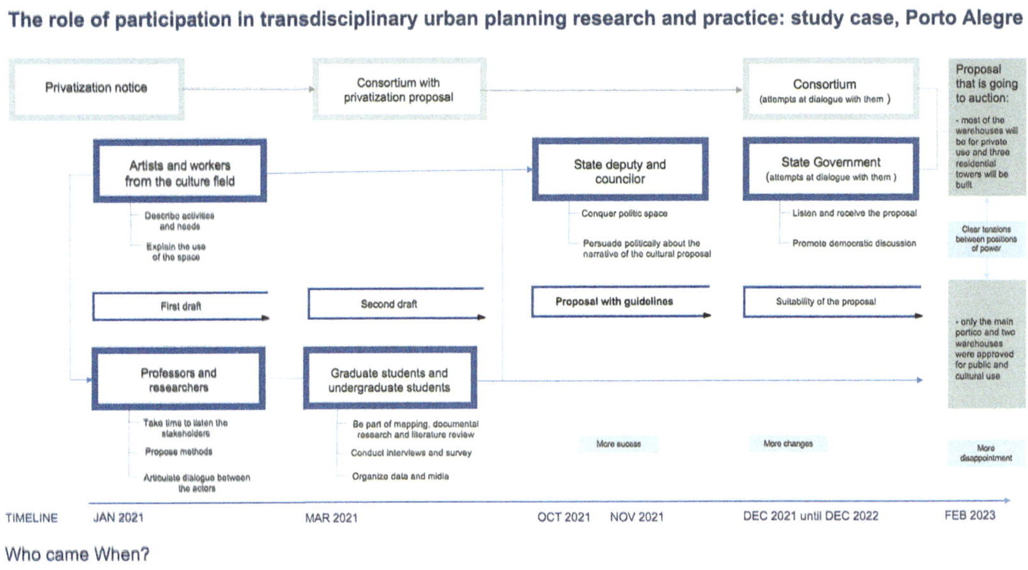

The role of participation in transdisciplinary urban planning research and practice: study case, Porto Alegre

Fig. 3.3 Conceptual model of the transdisciplinary research process created from the case study. (Source: authors)

Fig. 3.4 Cultural gathering at the pier with artists' collectives in September 2021 (© Custódio, Craidy Simões, 2021)

myriad contexts, including education, research or practical implementation. This fluidity often blurs the lines between research, practice and education projects, underscoring spatial methods as a unifying force that bridges the worlds of academia and practical urban planning.

The product of the interdisciplinary research and urban planning practice plays a crucial role, especially in disseminating the proposal to the wider public and validating and legitimising the guidelines for the harbour occupation. They demonstrate the feasibility of the proposal developed by the academic and cultural community while highlighting the conflict between the policy of concession/sale of public spaces, as proposed by the state government for the site, and the goal of ensuring universal and unrestricted access, materialising the public interest. The project enabled

Fig. 3.5 Meetings between the different actors, such as public hearings and negotiations with the state government, legislators as well as the Consortium, spanning from December 2021 to December 2022 (© Lersch, Custódio, 2021)

Table 3.2 Steps in the transdisciplinary case study Porto Alegre, Brazil

I. Preparation phase (January and March 2021)

Selecting the case

The initiative for this study originated from the artists' collective. The mobilisation of groups of artists and culture workers gained momentum as concerns about the potential privatisation of the pier grew. In January 2021, shortly after the collective's formation, artists reached out to university professors to propose a collaborative effort. The project was jointly spearheaded by theatre director Tania Farias (outside academia) and lawyer Jacqueline Custódio, along with Prof. Lersch, Prof. Marzulo, Prof. Costa and Prof. Fedozzi (co-leaderships) from the university. Efforts were additionally made to involve political representatives who could elevate the debate to parliamentary forums.

Searching for partners

The inaugural meeting between members of the collective and university researchers occurred in January 2021. We also sought to dialogue with representatives of the cultural community, social movements and the association of the Historical Centre neighbourhood.

Defining the problem

Upon analysing the proposal to transfer the public space on the pier to the private sector, it became evident that the place would lose many of its distinctive characteristics. If it were exclusively designated for commerce, services and real estate speculation, it would cease to be accessible to a diverse range of users.

Planning and writing project plan

In this step, we proposed formalising the '*Cais do Porto Cultural*' outreach project between the collective and the university.

II. Core phase (April 2021 to April 2023)

Meeting people

Dialogue with the cultural community. The experiences of the Condominium Scenic São Pedro in Porto Alegre (Brazil), the waterfront of Rosario (Argentina) and Matadero Madrid (Spain) were inspirations for the proposal of a pier with cultural activities, as expressed in the first meeting. We employed methods such as focus groups, discussion meetings and interviews. Around 30 artists participated in these initiatives.

Data survey and analysis

Between March and May 2021, a questionnaire/survey was developed in which the various cultural segments expressed their characteristics and needs in terms of facilities and spaces. Around 37 groups of artists participated in these initiatives, and we conducted discussions that covered circa 15 different types of artistic expressions (circus, theatre, music, visual arts, cinema, etc.). A comprehensive mapping of the quay site was additionally conducted during this period.

Discuss analysis

Between May and June 2021, utilising the information gathered from the questionnaires, we systematically organised the data, allowing for the identification of the specific needs of each group, definition of spaces and necessary infrastructure. This allowed us to develop the initial draft of the proposal. Two members of the collective played a crucial role as a bridge with the university team, consisting of four researchers and six students.

(continued)

Table 3.2 (continued)

Review progress	
Between June and October 2021, the guidelines and the initial draft were presented to the cultural community for their evaluation and suggestions. We employed methods such as focus groups and discussion meetings. Around 30 artists and workers from the cultural sector (stakeholders), 4 researchers and 6 students worked together.	
Moment of empowerment (1)	
On 11 September 2021, in collaboration with artists (approximately 60 artists and cultural workers), we organised an artistic parade and a cultural meeting at the pier. This event served as a demonstration of the significance of occupying public spaces with art and culture.	
Constructing scenarios	
Between June and October 2021, a series of online meetings were conducted to discuss and finalise spatial and management models. We employed methods such as focus groups and discussion meetings. Two members of the collective played a crucial role as a bridge with the university team comprising four researchers and six students.	
Moment of empowerment (2)	
28 October 2021—Installation of the Parliamentary Front in Defence of the Porto Alegre Pier; one state legislator and a councillor.	
Writing the proposal	
Between September and October 2021, the writing of the text and the production of the graphic elements were undertaken by a team comprising four researchers and six students. A literature review was also conducted during this period.	
Discussing scenarios	
Between December 2021 and December 2022, negotiations with government officials and legislators, including public hearings. This process involved the participation of two members of the collective and four researchers.	
Moment of empowerment (3)	
23 November 2021, presentation of the guidelines and the proposal at a press conference; two members of the collective and four researchers.	
Review progress	
Between January 2023 and February 2023, partial approval of the proposal, securing the retention of two warehouses and the portico as public spaces designated for the cultural area. Two members of the collective and four researchers.	
Discussing the final results	
Between February and March 2023, only the central portico and two warehouses out of the 11 were approved for cultural use, while the remaining structures were set to be auctioned. We organised a discussion meeting. Around 30 artists and workers from the cultural sector (stakeholders), 4 researchers and 6 students worked together.	
Final review	
Between March and April 2023, material organisation, literature review; four members of the collective and four researchers.	
III. Follow-up phase (April 2023 to December 2023)	
Drawing conclusions	
Between April and May 2023, from the attained results we evaluated the accomplishments and setbacks of the project implementation, as well as the urban impacts; literature review; four members of the collective and four researchers.	
Writing the paper	
Between May and June 2023, literature review; four members of the collective and four researchers.	
Media work	
In this final phase, seeking to publicise the actions.	
Implementing the follow-up process	
Between July and December 2023, waiting for the auction.	

the establishment of communication channels between academia and the cultural community, fortifying the ties between community representatives and segmented groups while identifying potential leaders. The capacity and active engagement in dialogue among these segments and the public, with a particular focus on the performing arts and visual arts, were crucial in integrating the issue of concession and the proposal for occupation into the daily lives of people.

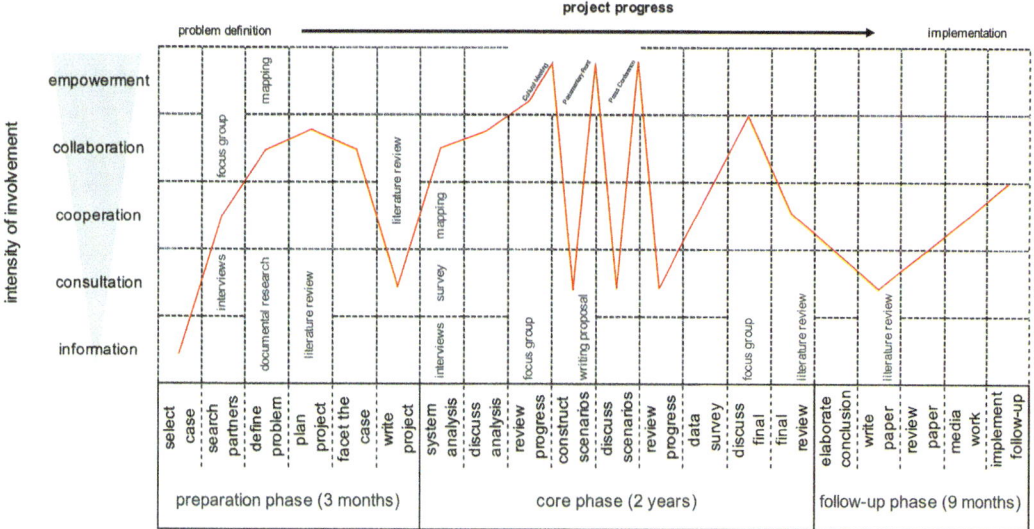

Fig. 3.6 Varying degrees of involvement and selection of applied techniques in the case study of Porto Alegre, Brazil

The achievement or non-achievement of the expected outcome hinges on numerous variables, including the ability to rally people around the cause, sustained group cohesion, dialogue skills among leaders, clarity of proposals, democratic participation of stakeholders, the leadership's analytical acumen regarding the proposal, the level of connection between the population and the disputed urban area, administrative structures and procedures that facilitate dialogue between the public and the government, and the political forces involved. In the case of the waterfront, the primary limitations were predominantly political in nature.

Although the main limitations stem from a global context that seeks to transform cities into market assets, the exercise of bridging interdisciplinary research and practice may serve as a model for other demands and initiatives, perhaps on a more localised scale in neighbourhoods or specific regions. Such future efforts can adapt the tactics used in this case to address unique challenges and contexts.

The employed methodology yielded positive results. One of the most important results was the pivotal role played by civil society in shaping the proposal. The collaborative interaction between the community and academia produced results that extended beyond the official forms of academic records. The formulation of the guidelines not only served as negotiating tool with government authorities but also as a tangible demonstration of the feasibility of the community's vision for the waterfront space. Artistic events were additionally organised to amplify the reach of information regarding the government's initiatives and the proposal advocated by civil society for the public and the democratic use of the waterfront area.

In the realm of administration, a proposal was advanced for a novel public asset management tool that would foster collaboration among the government, civil society and the private sector. In the sociological sphere, contributions aimed to understand how to navigate the numerous internal relationships established within the community group, among themselves and with academia, as well as external relationships with the government, media and the general population. The contributions from the cultural sector were also significant, as cultural creators laid the foundation for defining the outreach projects. Culture, in its broad spectrum, wields considerable communicative power in society, and the properties on the waterfront are recognised as historic and culturally relevant, thus legally protected under preservation norms that must be followed. All proposals were forwarded to the state government for inclusion in the bidding document to ensure their subsequent implementation.

3.5 Conclusion: Transdisciplinary and Participatory Integration Between Planning Pedagogy, Research and Practice

Operating within a transdisciplinary framework requires us to move beyond the confines of specialised disciplines. In the presented case study, we embarked on a collective exploration of Porto Alegre, focusing our attention on a distinct urban locale that resonated with various stakeholders. Our collaborative effort encompassed architects, urban planners, sociologists, administrators from the public policy domain and legal experts, reflecting the intricate nature of our subject of study. Our objective was to identify urban spaces ripe for activation and occupation, driven by the needs and aspirations of the communities involved.

Our investigation led us to identify open and unutilised spaces surrounding the pier, offering potential venues for open-air cultural events, including street theatre and circus performances. Furthermore, we recognised the untapped potential of the warehouses as enclosed spaces suitable for many activities, ranging from artistic performances and rehearsals to workshops, studios, services and administrative functions. This spatial exploration was informed by various methods such as diagrammatic sketching, interviews with stakeholders, questionnaires, surveys, mapping, drawing, photography and on-site visits, which helped us assess the possibilities and limitations of these spaces. Our experience reveals that transdisciplinary research and practice can yield formats and products that extend beyond the realms of social sciences and planning science, adapting to the context and expertise of the involved actors and ultimately allowing us to truly challenge the very boundaries between planning research and practice.

This case study does exhibit the hallmarks of transdisciplinary research as much as it aligns with the goals of SDG11, characterised by:

- Commitment to socially relevant issues. Our mission to transform a crucial urban area, the port quayside, into a public and democratic space in the face of looming privatisation threats underscores the social significance of our endeavour. We actively engaged non-academic stakeholders from culture, arts and organised civil society.
- The transcendence and integration of disciplinary boundaries: our research cohort comprised experts from diverse fields, including architecture, urbanism, sociology, law, management and public policy, as well as visual and performing artists. We embraced a wide array of discourses rooted in social practices, representing expressions of citizenship.
- Participatory research at its core: our process placed a premium on involving artists and cultural workers, ensuring that decisions were founded on thorough data collection regarding the needs, desires and demands of various social groups. While collaboration with the *Cais Cultural Já* collective was instrumental, it was not devoid of internal disputes and occasional tensions among researchers – a testament to the intricate dynamics at play.
- The quest for knowledge unity: our pursuit of a solution that catered to the needs of artists, cultural workers and the wider public, offering access to concerts, performances, studios, restaurants and open areas by the lake, solidified our vision for a public and democratic urban space. The collaborative stages, from framing the problem of ensuring public use of the Porto Alegre Pier to designing the proposal, were shaped through stakeholder dialogues.

Although the academic group played a more central role, given their technical expertise and the constraints imposed by the COVID-19 pandemic, significant strides were made in co-producing solutions that seamlessly intertwined transdisciplinarity and participatory research. It is safe to assert that in this context, the two concepts are synonymous and enriched by the diverse spatial methods we employed to inform our decision-making processes and to shape the final formats and products. This research provides a foundation for future endeavours, such as advancing the discussion of participatory research methods and the application of the methodology in other similar case studies as well as developing it further in the scholarship of teaching and research.

Acknowledgements We would like to express our sincere appreciation to SMUS alumni Dr. Megha Tyagi for her significant contribution to this chapter. Her careful reading and valuable insights into one of the evolving versions have been instrumental in shaping the content. We would also like to thank for the Cais Cultural team for their support during the outreach project.

References

Alves, A. (2005). A construção do Porto de Porto Alegre (1895–1930): modernidade urbanística como suporte de um projeto de estado. https://lume.ufrgs.br/handle/10183/5135

Argyris, C., & Schön, D. A. (1989). Participatory action research and action science compared: A commentary. *American Behavioral Scientist, 32*(5), 612–623. https://doi.org/10.1177/0002764289032005008

Barnerjee, N. (2021). Community-driven development as a mechanism for realizing global development goals. In *Better spending for localizing global sustainable development goals* (pp. 135–148).

Barton, H. (2010). *Shaping neighbourhoods: For local health and global sustainability* (2nd ed.). Routledge.

Bergmann, M. (2010). *Methoden transdisziplinärer Forschung: ein Überblick mit Anwendungsbeispielen.* Campus Verlag.

Bergold, J., & Thomas, S. (2012, January). Participatory research methods: A methodological approach in motion. *Forum Qualitative Sozialforschung / Forum: Qualitative Social Research 13*(1). Participatory Qualitative Research. https://doi.org/10.17169/FQS-13.1.1801

Berry, H. A., & Chisholm, L. A. (1999). *Service-learning in higher education around the world: An initial look.* International Partnership for Service-Learning.

Brandt, P., Ernst, A., Gralla, F., Luederitz, C., Lang, D. J., Newig, J., Reinert, F., Abson, D. J., & von Wehrden, H. (2013). A review of transdisciplinary research in sustainability science. *Ecological Economics, 92*(August), 1–15. https://doi.org/10.1016/j.ecolecon.2013.04.008

Croese, S., Dominique, M., & Raimundo, I. M. (2021). Co-producing urban knowledge in Angola and Mozambique: Towards meeting SDG 11. *Npj Urban Sustainability, 1*(1), 8. https://doi.org/10.1038/s42949-020-00006-6

Curdes, G. (1995). *Stadtstrukturelles Entwerfen.* W. Kohlhammer.

Fals-Borda, O. (1987). The application of participatory action-research in Latin America. *International Sociology, 2*(4), 329–347. https://doi.org/10.1177/026858098700200401

Fischer, F., & Forester, J. (Eds.). (1993). *The argumentative turn in policy analysis and planning.* Duke University Press. https://tu-berlin.hosted.exlibrisgroup.com/primo-explore/search?query=any,contains,9921985160402884&tab=tub_all&search_scope=TUB_LOCAL&vid=TUB&lang=de_DE&offset=0

Forester, J. (1999). *The deliberative practitioner: Encouraging participatory planning processes.* MIT Press. https://tu-berlin.hosted.exlibrisgroup.com/primo-explore/search?query=any,contains,9921985159802884&tab=tub_all&search_scope=TUB_LOCAL&vid=TUB&lang=de_DE&offset=0

Fritz, L., & Binder, C. (2018). Participation as relational space: A critical approach to Analysing participation in sustainability research. *Sustainability, 10*(8), 2853. https://doi.org/10.3390/su10082853

Gehl, J. (1987). *Life between buildings: Using public space.* Van Nostrand Reinhold.

Giseke, U. (2021). *Urban design methods: Integrated urban research tools.*

Healey, P. (1993). Planning through debate: The communicative in planning theory. In F. von Fischer & J. Forester (Eds.), *The Argumentative Turn in Policy Analysis and Planning* (pp. 233–253). Duke University Press.

Healey, P. (1997). *Collaborative planning: Shaping places in fragmented societies.* UBC Press.

Heinrich, A. J., Marguin, S., Million, A., & Stollmann, J. (2021). *Handbuch qualitative und visuelle Methoden der Raumforschung.* utb – transcript.

Ibge (2022). Instituto Brasileiro de Geografia e Estatística. *Censo Demográfico 2022.* Rio de Janeiro: IBGE. https://www.ibge.gov.br/

Kotval, Z. (2003). Teaching experiential learning in the urban planning curriculum. *Journal of Geography in Higher Education, 27*(3), 297–308. https://doi.org/10.1080/0309826032000145061

Lang, D. J., Wiek, A., Bergmann, M., Stauffacher, M., Martens, P., Moll, P., Swilling, M., & Thomas, C. J. (2012). Transdisciplinary research in sustainability science: Practice, principles, and challenges. *Sustainability Science, 7*(S1), 25–43. https://doi.org/10.1007/s11625-011-0149-x

Lawson, B. (2009). *What designers know*. Repr. Elsevier/Architectural Press.

Million, A. (2021). Mental Maps und narrative Landkarten. In A. J. von Heinrich, S. Marguin, A. Million, & J. Stollmann (Eds.), *Handbuch qualitative und visuelle Methoden der Raumforschung* (pp. 293–308). utb – transcript.

Mobjörk, M. (2010). Consulting versus participatory Transdisciplinarity: A refined classification of transdisciplinary research. *Futures, 42*(8), 866–873. https://doi.org/10.1016/j.futures.2010.03.003

Marzulo, E. P., Lersch, I. M., Fedozzi, L., & Costa, P. A. (2021). *Cais Cultural*: Proposta de Ocupação Cais do Porto de Porto Alegre. https://propostacaisdoportoalegre.blogspot.com/p/diretrizes-gerais-cais-cultural.html

Pohl, C. (2010). From Transdisciplinarity to transdisciplinary research. *Transdisciplinary Journal of Engineering & Science, 1*(1). https://doi.org/10.22545/2010/0006

Reason, P., & Bradbury, H. (Eds.). (2008). *The Sage handbook of action research: Participative inquiry and practice* (2nd ed.). SAGE Publications.

Sara, R. (2011). Learning from life – Exploring the potential of live projects in higher education. *Journal for Education in the Built Environment, 6*(2), 8–25. https://doi.org/10.11120/jebe.2011.06020008

Sara, R., & Jones, M. (2018). The university as agent of change in the city: Co-creation of live community architecture. *International Journal of Architectural Research: ArchNet-IJAR, 12*(1), 326–337.

Schäpke, N. (2017). Reallabore im Kontext transformativer Forschung: Ansatzpunkte zur Konzeption und Einbettung in den internationalen Forschungsstand.

Scholz, R. W., & Steiner, G. (2015). Transdisciplinarity at the crossroads. *Sustainability Science, 10*(4), 521–526. https://doi.org/10.1007/s11625-015-0338-0

Smit, S., Musango, J. K., & Brent, A. C. (2019). Understanding electricity legitimacy dynamics in an urban informal settlement in South Africa: A community based system dynamics approach. *Energy for Sustainable Development, 49*(April), 39–52. https://doi.org/10.1016/j.esd.2019.01.004

Souza, C. F., & Müller, D. M. (2007). *Porto Alegre e sua Evolução Urbana*. 2. ed. Porto Alegre, Editora da Universidade.

Stauffacher, M., Flüeler, T., Krütli, P., & Scholz, R. W. (2008). Analytic and dynamic approach to collaboration: A transdisciplinary case study on sustainable landscape development in a Swiss Prealpine Region. *Systemic Practice and Action Research, 21*(6), 409–422. https://doi.org/10.1007/s11213-008-9107-7

Streck, D. B., & Rodrigues Brandão, C. (Eds.). (2005). Participatory action research in Latin America. *International Journal of Action Research, 1*(1), 43–68.

UN, United Nation (2015). Transforming our world: the 2030 Agenda for Sustainable Development. https://sdgs.un.org/2030agenda

Watson, V. (2014). Co-production and collaboration in planning – The difference. *Planning Theory & Practice, 15*(1), 62–76. https://doi.org/10.1080/14649357.2013.866266

Watson, V. (2016). Shifting approaches to planning theory: Global north and south. *Urban Planning, 1*(4), 32–41. https://doi.org/10.17645/up.v1i4.727

Watson, V., & Odendaal, N. (2013). Changing planning education in Africa: The role of the Association of African Planning Schools. *Journal of Planning Education and Research, 33*(1), 96–107. https://doi.org/10.1177/0739456X12452308

Weyh, C. B., & Streck, D. B. (2003). Participatory budget in Southern Brazil: A collective and democratic experience. *Concepts and Transformation, 8*(1), 25–42. https://doi.org/10.1075/cat.8.1.03wey

Yiftachel, O., & Huxley, M. (2000). Debating dominance and relevance: Notes on the 'communicative turn' in planning theory. *International Journal of Urban and Regional Research, 24*(4), 907–913.

Zscheischler, J., Rogga, S., & Weith, T. (2014). Experiences with transdisciplinary research: Sustainable land management third year status conference. *Systems Research and Behavioral Science, 31*(6), 751–756. https://doi.org/10.1002/sres.2274

Prof. Dr. Inês Martina Lersch is Professor of Urban Planning at the Faculty of Architecture of the Federal University of Rio Grande do Sul. She has a BA in Architecture and Urban Planning (1998) from the same University, where she also completed her MA in Civil Engineering (2003) and her PhD in Urban and Regional Planning (2014). Her work primarily focuses on urban planning, planning history and heritage at the Postgraduate Program in Urban and Regional Planning (PROPUR/UFRGS). She is a SMUS Post-Doctoral Researcher (2022) and also a SMUS Partner.

Prof. Dr. Angela Million, née Uttke, is Professor for Urban Design and Urban Development at TU Berlin, Germany. She is Director of SMUS. Her most current research explores educational landscapes, neurourbanism, multifunctional infrastructure as well as hybrid spatial constructions of young people within the Collaborative Research Center 1265 'Refiguration of Spaces'. She is founding member of JAS (Jugend Architektur Stadt e.V.), a non-profit association dedicated to built-environment education for young people. As an urban planner and urban designer, she is part of design juries and councils in different German cities.

Jacqueline Custódio is a lawyer who specialises in Public Law and a member of ICOMOS Brazil. She holds an MA in Museology and Heritage from the Federal University of Rio Grande do Sul, where she is currently pursuing her PhD in Urban and Regional Planning (PROPUR/UFRGS) with a focus on participatory planning and cultural heritage in the city of Porto Alegre. She is a member of the Sectorial Collegiate of Material Heritage at the National Council of Cultural Policy of Brazil's Ministry of Culture (CNPC/MinC). She worked as a member of both the Rio Grande do Sul State Council of Culture (CEC/RS) and the Sectorial Collegiate of Memory & Heritage of the Porto Alegre Municipal Council of Culture.

Open Access This chapter is licensed under the terms of the Creative Commons Attribution 4.0 International License (http://creativecommons.org/licenses/by/4.0/), which permits use, sharing, adaptation, distribution and reproduction in any medium or format, as long as you give appropriate credit to the original author(s) and the source, provide a link to the Creative Commons license and indicate if changes were made.

The images or other third party material in this chapter are included in the chapter's Creative Commons license, unless indicated otherwise in a credit line to the material. If material is not included in the chapter's Creative Commons license and your intended use is not permitted by statutory regulation or exceeds the permitted use, you will need to obtain permission directly from the copyright holder.

Part II
The Critical Role of Spatial Methods in Transdisciplinarity

4. Urban Sustainable Interactions by Homeless People Here and Now via Spatial Methods

Fraya Frehse, Caio Moraes Reis, and Giulia Pereira Patitucci

4.1 An Introduction to Homelessness and SDG11

Is it possible to envisage within the socially dramatic issue of homelessness any contributions for SDG11? This chapter delves into this somewhat counterintuitive question with the aid of a particular set of spatial methods, which was developed in the framework of a broader transdisciplinary, research-practice pilot project of the Global Center of Spatial Methods for Urban Sustainability, or SMUS (GCSMUS, 2023a). By encompassing two specific sets of qualitative research techniques sensitive to the social and relational dimension of space, the so-called SMUS Toolkit (GCSMUS, 2021) helped us inquire about the everyday life of men, women and children who dwelled in the streets and squares of the largest Latin American metropolis during the Covid-19 pandemic.

Drawing from the analytical and interpretive implications of the SMUS Toolkit, our statement is that these contributions are of an interactional-spatial nature. The contributions are (re)produced here and now: they concern daily patterns of face-to-face interaction by and around homeless people in public spaces. Regarding homelessness particularly in Covid-19 São Paulo, the contributions encompass, on the one hand, (i) environmentally sustainable and (ii) environmentally inclusive interactions by homeless people with material objects, animals and plants in streets and squares. On the other hand, they refer to (iii) a socially inclusive knowledge about homeless people by practitioners devoted to homelessness. While the first two types of interaction are already ongoing, present-day involuntary contributions of homeless people to the SDG11 targets 11.6 ('reduce the *environmental* impact of cities'; emphasis added) and 11.7 ('provide access', among others, 'to *inclusive* green and public spaces'; emphasis added), the third type features as a future possibility for practitioners to target 11.1 (ensure universal access to, among others, '*adequate*' housing with '*sufficient* living area'; emphasis added).

This chapter demonstrates this threefold statement in three steps. An initial conceptual section elucidates why the possibility of a collaborative connection between homelessness to SDG11 appears counterintuitive and why we chose to examine it using the SMUS Toolkit. The subse-

F. Frehse (✉) · C. M. Reis
Department of Sociology + SMUS, University of São Paulo, São Paulo, Brazil
e-mail: fraya@usp.br;
caio.moraes.reis@alumni.usp.br

G. P. Patitucci
Federal Ministry of Management and Innovation in Public Services, Brasília, Brazil

quent methodological section outlines and justifies the unique scholarly contribution of this transdisciplinary framework, which acts as a set of 'glasses' for examining homelessness in Covid-19-era São Paulo by means of their everyday spatialities. Then, it is time to apply the SMUS 'glasses' into analytical and interpretive action. The concluding section summarises six lessons of our spatial-methodological and transdisciplinary project for both academics and practitioners regarding urban sustainability—in and beyond the street.

4.2 A Counterintuitive Issue

The international academic community holds a virtually unanimous view that homelessness presents uncountable challenges and undeniable threats to the UN 2030 Agenda. Homelessness is understood as a phenomenon to be 'eliminated' (Farha, 2019), 'eradicated' (Morris, 2020), 'prevented' (Speak, 2013), 'ended' and 'fought against' (FEANTSA, 2023). Before the pandemic, the relationship between homelessness and the urban sustainability agenda was largely underexplored (for an exception see Salcedo, 2019). However, the vulnerability faced by homeless people globally amid the Covid pandemic has pushed forward initiatives to make homelessness an explicit concern of the SDGs (Casey & Stazen, 2021). In December 2021, the UN General Assembly created a set of resolutions that directly addressed homelessness as 'a condition where a person or household lacks safe habitable space, which may compromise their ability to enjoy social relations, and includes people living on the streets, in other open spaces or in buildings not intended for human habitation, people living in temporary accommodation or shelters for people experiencing homelessness, and, in accordance with national legislation, may include, among others, people living in severely inadequate accommodation without security of tenure and access to basic services' (UN, 2022: 4; see also Casey & Stazen, 2021: 68).

We argue that one reason for the common assumption that homelessness is incompatible with urban sustainability is the temporality implicit in the concept of 'sustainable development' that underpins the UN 2030 Agenda. Sustainability is an attribute of individual or collective 'initiatives'—ranging from (inter)actions to policies as well as social relationships and practices—that favour a *future*, balanced subsistence of the planet's environmental, economic and social resources for human and non-human generations to come.[1]

Against this backdrop, the everyday life of the urban poor—among which those living in homelessness are one extreme expression—may at first sight appear as widely unsustainable. Scholarship on urban poverty has consistently emphasised the structural role of *present-time* immediacy and improvisation in the everyday of poor people in cities. Terms commonly used to address this fleeting temporality range from 'survival strategies' for coping inclusively with state and street violence (for a recent overview see Deckard & Auyero, 2022), to the more recent 'endurance', which captures daily practices that 'blend in and exceed' the norm of mere survival (Simone, 2019). Not to forget 'social resilience', conceived as urban-systemic adaptations to 'chronic shocks' (Fahlberg et al., 2020), and the Brazilian '*viração*'—a native category coined to address both the 'identities' and 'rules of interaction' forged by street dwellers amidst their everyday strategies for getting-by in São Paulo's public spaces (Gregori, 2000; Frehse, 2013; see also Frehse, 2021a, b).

However, this body of literature also prolifically heightens the *future-driven* capacity of the poor to cope with the structural shocks that make up their urban everyday life. Simone (2019: 20) offers a poetic summary of this skill when addressing care among (poor) 'residents' as 'something that enables endurance, not necessarily their own endurance as human objects, but the endurance of care indifferent to whatever or whoever it embraces'.

---

[1] The historically groundbreaking UN Brundtland Report defines *sustainable* development as development 'that meets the needs of the present without compromising the ability of future generations to meet their own needs' (WCED, 1987: 43; see also UN, 2023).

This trait signals to the possibility that the everyday practices of the urban poor also bear an urban-sustainable dimension. Nevertheless, this remains underexplored in both the academic and public-political debate on the interplay between homelessness and urban sustainability. The UN Agenda addresses this kind of contribution in connection with social activities such as waste-picking/collecting (Dias, 2016; Gutberlet, 2021).

Considering this current state of academic and policy affairs, it may seem far-fetched and contradictory to explore the potential contributions of homelessness to SDG11. How can a phenomenon that poses undeniable social threats to urban sustainability be beneficial?

Our interest in this issue emerged in the framework of the aforementioned project, whose academic aim was to 'illuminate how spatial methods may enhance the relationship between homelessness and urban sustainability' (GCSMUS, 2023a). Initially pursuing interdisciplinary research objectives and subsequently striving for transdisciplinary research-practice goals with the aid of the SMUS Toolkit under the supervision of one of us (Frehse), eight graduate student-researchers and practitioners, including two of us (Reis and Patitucci)—from now on referred to as *our team*—mobilised an *ethnographic* set of spatial methods to address the *everyday spatialities of homelessness* in Covid-19 São Paulo (2020–2021). Embracing ethnography as an 'epistemological perspective' (Frehse, 2006: 300), our team simultaneously made strange what seemed familiar and familiar what seemed strange to us while engaging in face-to-face interaction with homeless people and practitioners involved in social work related to homelessness from November 2020 to April 2022. Our subject of ethnographic (de)familiarisation was the *everyday spatialities* of homelessness: that is, the bodily arrangements that men, women and children make on a daily basis of the public places where they dwell, while ascribing meanings to their own face-to-face interactions with people, institutions, objects, animals and plants therein (Frehse, 2022: 5; see also GCSMUS, 2023a).

The *interactional sensitivity* embedded in the 'ethnographic perspective' (Frehse, 2006: 300) empowers us to identify, within the 'condition' of homelessness, elements beyond the deficits and deficiencies that make up the aforementioned UN definition. Homelessness thus becomes one specific 'pattern of bodily use of urban public spaces' (Frehse, 2016: 135), which we define as the most physically, legally and informationally accessible places in cities, such as streets and squares (Frehse, 2018: 33). Homelessness 'concerns the regular physical permanence of human beings in streets, squares, and other public urban places for overnight stays and, thus, for dwelling (the etymology of the term "to dwell" comes from the Middle English *dwellen*: to physically delay, live, remain, persist)' (Frehse, 2020: 2). Moreover, while poor people are distinct from poverty itself, homelessness is a definite 'sociospatial characteristic of an urban society which produces and reproduces itself globally precisely by way of, among others, this phenomenon' (Frehse, 2020: 3; see also Harvey, 2012: 23, 35).

Based on the analytical and interpretive implications of the SMUS Toolkit, the following pages will bring to the conceptual forefront our three-fold statement regarding the contributions to SDG11 within the issue of homelessness. Beforehand, we must embark on our entire methodological journey. After all, our argument is the outcome of a comprehensive analytical and interpretative examination of the *ethnographic perspective* applied to data produced with the aid of one specific set of spatial methods, all within the framework of a single transdisciplinary initiative aimed at advancing urban sustainability.

4.3 Contextualising the SMUS Toolkit

Ethnographic research on homelessness has been commonplace in the social sciences since Nels Anderson's groundbreaking study on the Chicago 'hobos' (Anderson, [1923] 1967). However, Anderson only indirectly references ethnography. His introduction to the book's reissue mentions the co-supervision of Robert Park, who strongly advocated for sociologists to use anthropological methods similar to those of journalists—i.e., primarily based on firsthand and eye-witnessing 'observation'—to understand 'human behaviour in the urban environment' (Park, [1925] 1967: 3). Park advised Anderson to '[w]rite down only what you see, hear, and know, like a newspaper reporter' (Anderson, [1923] 1967: xii).

Much less common is the use of ethnography in research *and* professional practice with the transdisciplinary goal of bridging the gap between academia and practice (see, for example, Zscheischler et al., 2022; Von Schönfeld et al., 2023). Whether taken as an investigative method or as an epistemological perspective, ethnography is more typically associated with research than to research-practice (note the hyphen!), as in the forementioned SMUS project.

A similar trend applies to ethnography as a *spatial* research method. Research projects are commonplace (for example Genz & Tschoeppe, 2021; Low, 2017; Wetzels, 2021), whereas transdisciplinary endeavours are scarce—and even scarcer when it comes to the UN urban sustainability Agenda (for exceptions see Kagan, 2019; Dymitrow & Ingelhag, 2019; Low, 2023; and Chap. 6 in this book).

Against the backdrop of this debate, we argue that, when taken particularly as an epistemological perspective, ethnography has an operational versatility that renders it unique for transdisciplinary projects geared towards urban sustainability. Therefore, our model case is precisely the four-step methodological path—i.e., *method*, from the Greek "pursuit of a way"—pursued in the framework of the SMUS pilot project (Fig. 4.1):

During the project's Phase 1, primarily focused on research objectives, the *scientific* versatility of the ethnographic perspective was decisive. With its epistemological aid, we empirically inquired into the everyday spatialities of homelessness. This implied employing research techniques typical of ethnographic research: participant observation, semi-structured interviews, and visual tools (such as the production of sketches, photographs, drawings, and maps). This 'Data Collection' phase (Fig. 4.1) was a three-month interdisciplinary Training Program on data collection and spatial methods (led by Frehse and Castillo Ulloa) and attended by the aforementioned eight graduate students (architecture, urban planning, sociology, anthropology, history, social psychology and nursing), some of which also practitioners working on homelessness (particularly Patitucci alongside the team's other architect, a social psychologist and a nurse). The student–researcher–practitioner team shared both research and practice interests in the everyday of homelessness in São Paulo. From November 2020 to January 2021, the team learned to put the ethnographic perspective into empirical action.

The subsequent analytical project Phase 2 (Fig. 4.1) comprised the qualitative analysis of the collated text and visual information in search of clues about the everyday spatialities of homelessness. From March to September 2021, two of us (Frehse and Reis) accomplished this task via the software MAXQDA. The qualitative analysis provided the foundation for Frehse to outline the twofold operational contours of the SMUS Toolkit for transdisciplinary research projects. As explained elsewhere (GCSMUS, 2023b), the Toolkit encompasses, on the one hand, a collection of *ethnographic* observation techniques about the everyday spatialities of the researched subjects (i.e., direct and participant observation, and go-along interviews). On the other hand, it comprises a set of *ethnographic* visualisation techniques of these same spatialities (i.e., mappings by means of sketches, photographs, drawings, etc., including audio recordings via WhatsApp and other social media; in short, any technical device that aids in visualising the everyday spatialities in question). In other words, the ethnographic perspective also underpins the spatial-methodological toolkit specifically designed for our team to address the relationship between homelessness and urban sustainability also within a transdisciplinary framework.

Therefore, it comes as no surprise that ethnography also grounded the project's subsequent, essentially transdisciplinary Phase 3 (Fig. 4.1). By delving into the *transdisciplinary* flexibility of ethnography, Frehse endowed the ethno-

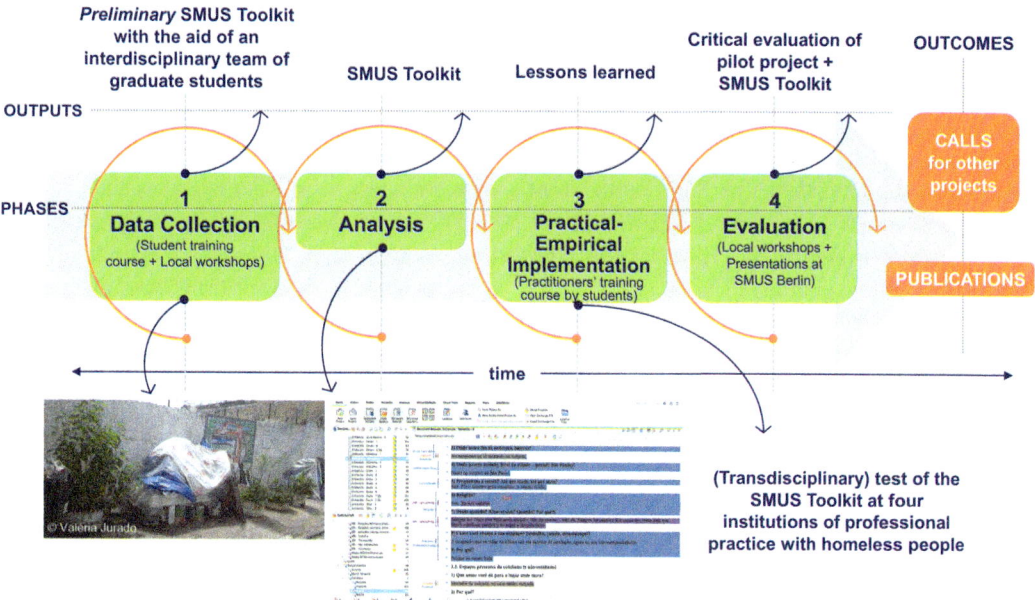

Fig. 4.1 The four-phase dynamics of the pilot project. (© SMUS Action 4 Team—GCMSUS, 2023a)

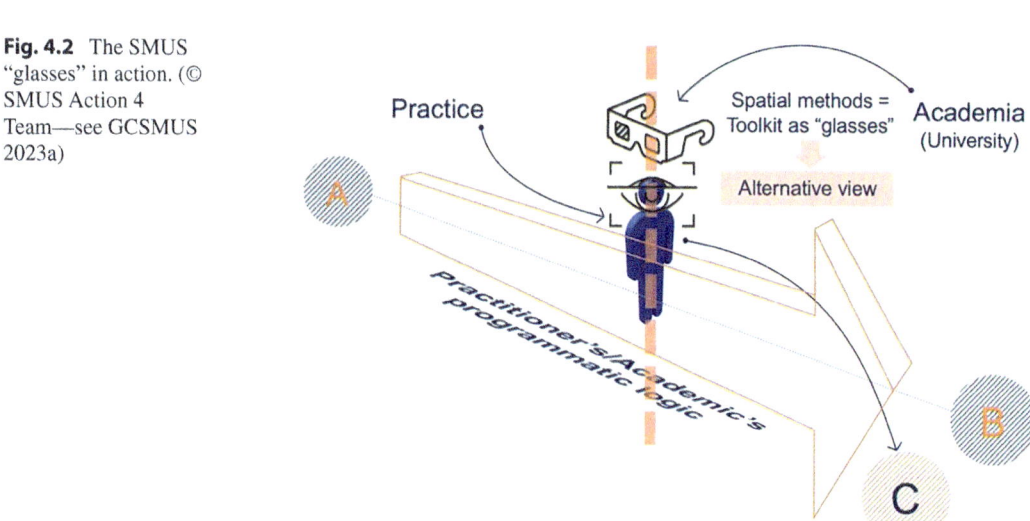

Fig. 4.2 The SMUS "glasses" in action. (© SMUS Action 4 Team—see GCSMUS 2023a)

graphic perspective with a scientific-communicational dressing. She framed the SMUS Toolkit as a metaphoric pair of 'glasses', which both academics and practitioners are invited to wear to gain an alternative view of their own research and practice issues (Fig. 4.2).

Since then, this schematic, visual representation of our spatial-methodological approach to urban sustainability has become essential for conveying its underlying ethnographic perspective with a wider transdisciplinary audiences. Therefore, the *pedagogic* versatility of ethnography as an epistemological perspective could enter the scene.

Although practitioners were involved in the first phase, the full transdisciplinary nature of the SMUS project evolved during its third phase. Between October and December 2021, the previously trained student–researcher–practitioner team implemented the SMUS Toolkit in practical-empirical terms (Fig. 4.1) under the supervision of Frehse. Our pedagogic strategy was to conduct a 4-week Training Course on 'Spatial Methods for Professional Practice with Homelessness' to 26 practitioners based at four major institutions that tackle homelessness in São Paulo from within–given their geographical location in the city's downtown area, which incidentally concentrates homeless people in spatial terms (60%; Qualitest, 2021a). The institutions are: (i) a CBO named State Movement of the São Paulo Population in Street Situation (MEPSRSP); (ii) the Special Social Approach Service (SEAS), spearheaded by the City and implemented by an NGO, responsible for identifying homeless people in the city and integrating them into the City's social-service network; (iii) the São Martinho Living Centre, a spatially stationary social-assistance hub managed by the City and operated by an NGO; (iv) the spatially mobile axis of that same NGO, i.e., the also City-led Street Clinic (*Consultório na Rua*), which focuses on the physical and mental healthcare of homeless people approached directly in the streets. At the start of Phase 3 of the project, our team was already acquainted with these institutions. The ethnographic training during the previous Phase 1 had taken place in the framework of our interactions with them (see Frehse, 2020).

Still via the ethnographic perspective, we creatively explored the pedagogic dimension of the SMUS Toolkit in the here and now of situations of bodily and materially mediated, face-to-face interaction with the practitioners of those four institutions. Before the course began, our team addressed its major pedagogic challenge: How can we use the SMUS Toolkit to empower social workers to create a transdisciplinary difference in their professional practice without attempting to redefine them as academics? Based on the ethnographic lessons learned from the project's previous phases, we drew on the 'problem-posing education' developed by the internationally renowned Brazilian educator Paulo Freire (1967; Bartlett, 2008). We converted the previously depicted, everyday spatialities of homelessness into eight conversational topics that pairs of student–researchers–practitioners proposed to the practitioners from the four institutions who attended the course. Our motto was: 'We invite you to put on the SMUS "glasses" so we can establish a dialogue about the topics at hand, while we accompany you in your work routine with homeless people in the streets and squares'.

The immersion took place in the situational framework of eight 'exchange meetings'. During those situations, the four-team duos encouraged the practitioners to exchange their verbal and visual impressions with them verbally, visually or textually regarding the following issues: (i) the practitioners' conceptions about homeless people prior to the start of the course; (ii) the daily routine of homeless people within the confines and beyond the walls of the practitioners' institutions; (iii) the spatialities of the past, present and future dwellings of homeless people; (iv) the places where homeless people circulate in their daily lives (streets, squares, home, the institution at stake, shelters, tents, etc.); (v) the role of violence; (vi) the role of personal objects, pets, plants, and friends and family; (vii) how homeless people engage in leisure activities; and (viii) the practitioners' definitions of homeless people after the course. Hence, each physically mobile and/or immobile exchange meeting became a situation where our student-researcher-practitioner team and the practitioners 'made strange' their own (pre-)conceptions regarding homeless people before and after the SMUS training courses they attended.

The project's Phase 4 (Fig. 4.1) was evaluative. It focused on the practical-empirical results attained through the ethnographic exchange of gazes by means of the SMUS 'glasses'. In April 2022, a group of academics and practitioners, directly or indirectly involved in the four stages of the project, congregated for a critical discussion on the potentials and constraints of applying the SMUS Toolkit to bridge the gap between academia and practice regarding homelessness and its consequences for urban sustainability (GCSMUS, 2023c).

The implications of this comprehensive, spatial-methodologically driven, transdisciplinary process reach far beyond the seven research and practice outputs generated to date

(GCSMUS, 2022). The research-practice team produced a collection of ethnographic data sources, which comprise (i) eight field diaries by each student-researcher-practitioner and/or the duos about the project's phases 1 and 3 and (ii) the transcriptions of 29 semi-structured interviews conducted during Phase 1 with homeless people about their everyday spatialities. The first batch of diaries encompasses 776 pages of written content alongside sketches, photographs, drawings, maps and WhatsApp videos that reference the everyday spatialities of homeless people in city streets, squares and institutional sites from November 2020 to January 2021. In turn, the second set comprises 465 pages with a slightly varied content produced between November and December 2021: rather than WhatsApp videos, text and audio messages produced by the practitioners who attended the Phase-3 Training Course. We proactively encouraged the production of these variegated types of materials. Our goal was to diversify our means of communication with the practitioners as much as possible to capture the qualitative changes in their prior knowledge about the everyday spatialities of homelessness.

The field diaries are particularly vital for addressing the counterintuitive issue that constitutes the focus of this paper. As we will see moving forward, they reveal the *analytical and interpretive* versatility of the ethnographic perspective that underlies the SMUS Toolkit.

4.4 Approaching Urban-Sustainable Homelessness via the SMUS 'Glasses'

In this dataset, we have a temporal sequence of 237 ethnographic field reports and 596 audio–visual materials.[2] This material holds clues about the ethnographically induced, non-verbal and verbal *intercations* of our team with homeless people (Phase 1) and with practitioners who worked with homeless people (Phase 3) in the streets, squares, and sidewalks of São Paulo, during a timespan in which the structurally improvised routines of these people were challenged by conjunctural socioeconomic and health vulnerabilities brought about by the pandemic.

Brazil has been severely affected by Covid-19. Upon the completion of the SMUS pilot project in April 2022, the nation had recorded circa 663,5000 deaths amid almost 30,5 million officially confirmed infections (Ministério, 2022, 2023). However, the highest daily death toll (4200) stems from 1 year before. São Paulo, in particular, experienced a historical event in homelessness from 2019 to 2021: a nearly 29% rise in family homelessness (Qualitest, 2021a: 25). The pandemic has exacerbated the then ongoing (especially since 2019) social-political dismantlement of Brazil's recent, albeit far-reaching, social assistance structure spanning the municipal, regional and federal levels. These public-political circumstances required from our team an extraordinary combination of humanitarian empathy and readiness, which the ethnographic perspective helped us reinforce.

For the purposes of this chapter, the ninefold set of everyday spatialities of homelessness, which emerged during the analytical phase of the project, holds significant importance both in analytical and interpretive terms. Forged methodologically with the aid of the ethnographic perspective ingrained in the SMUS Toolkit, these spatialities contain a face-to-face, interactional dimension.

First, let us conceptualise these spatialities via Goffman's bodily and materially mediated approach to the situational immediacy of social interaction (Frehse, 2021a: 50). From this standpoint, the public spaces in São Paulo where homeless people congregate are depicted in the field diaries as places of (i) strict and very rigid daily routines of (non-)verbal interactions that secure these peoples' physical survival and socioeconomic subsistence; (ii) everyday interactions implicit in their bodily use of these spaces for dwelling; (iii) work-

---

[2] Regarding the written reports, there are 114 accounts from Phase 1 and 123 notes from Phase 3. The 325 audio-visual sources from Phase 1 comprise 51 sketches, 222 photos, 19 drawings, 23 maps, and 10 WhatsApp videos; in turn, Phase 3 resulted in 17 sketches, 196 photographs, 24 drawings, 16 maps, and 18 WhatsApp audio messages, which totals 271 materials.

related interactions implicit in activities ranging from begging to waste-picking for recycling; (iv) the bodily and materially mediated circulations of homeless people thru and fro shelters and social-assistance institutions; (v) physically and/or symbolically violent interactions of homeless people with peers and third parties; (vi) their affectionate interactions with pets; (vii) their affectionate interactions with plants; (viii) family-based interactions with peers; and (ix) playful interactions with peers and third parties. In a nutshell, these spatialities provide insights into the 'spatialisation patterns of social interaction' (Frehse, 2021a: 50).

Sociologically, the concept refers to symbolic regularities (i.e., rules) involved in the temporal immediacy of bodily and materially mediated human (non-)verbal, face-to-face interaction with one another, with other living beings and/or material and symbolic goods in places within the spatial boundaries that make up 'situations' of face-to-face interaction (Goffman, 1963: 18; see also Frehse, 2021a: 50). The spatialisation patterns were not only staged by homeless people and practitioners in front of our team in the public spaces of the pandemic city within the spatial boundaries of SMUS Toolkit-induced situations of face-to-face interaction. They also took on a spatial dimension within the materiality of the team's field diaries, more specifically in the textual and visual references to the everyday spatialities of homelessness in Covid-19 São Paulo.

Based on this rationale, the nine forementioned spatialities disclose three types of spatialised interaction; or, better stated, of *urban sustainable* face-to-face interchanges *by* and *about* homeless people in public spaces. Two of these are *already ongoing* contributions while the third is a *future, possible* contribution involving the application of the SMUS Toolkit to address homelessness in the context of SDG11.

4.4.1 Present-Day, Environmentally Sustainable and Inclusive Public Spaces

Three of the nine aforementioned spatialities are particularly revealing of ongoing, environmentally (i) sustainable and (ii) inclusive social interactions by homeless people in Covid-19 São Paulo. Therefore, we establish a dialogue with the UN definition of environment as 'physical environment or biota' that encompasses 'natural and managed ecosystems' (UN, 1992: 162). While this definition remains alien to the social-relational dimension of space that underlies our project, it does draw attention to the fact that the aspiration of urban sustainability is inseparable from initiatives attentive to the non-human world in cities in a broad sense. Consequently, not only the collection of solid waste (see Target 11.6) by homeless people but also their care for animals and plants can be viewed in a new light.

Within the set of 27 fieldnotes, 17 photographs and 5 sketches with (non-)verbal references to homeless people involved in environmental issues during their face-to-face interchange with members of our team, almost half of these notes (i.e., 11) may be analytically collated around one specific everyday spatiality. Public spaces are the daily workplaces for a group of (ten) cisgender men and one transgender woman who are involved in recycling solid waste, which the diary authors usually named 'recyclable materials'—and only once designated 'rubbish' and 'small cans'.

The most significant variation in the data collection is the referential spatiality of the face-to-face interactions around recycling work. Three fieldnotes alluded to 'squares', which hosted gatherings of tents and/or handcarts belonging to homeless people. Patitucci, for example, noticed in January 2021 that the transgender woman 'Eva[3] had a small table in front of her tent' at the historic downtown square where she lived. There 'she piled up the recycled materials she had managed to collect' (Field diary [FD] Patitucci 2020–21: 68 regarding 8.01.2021). The other eight cases concern 'the city' in general. During that same month, our team member Ednan Santos noticed that 'Valter's work [with "recyclable materials"] takes him far away' geographically from the square where he sleeps rough, albeit still

---

[3] To protect the privacy of the homeless people mentioned, we have replaced their real names with pseudonyms.

in the downtown area. For his work purposes, 'he doesn't plan any route in advance; he simply circulates through the city'. Hence, 'I sometimes sleep in the street because I'm far away from home. Sometimes I don't even know where I am' (FD Santos 2020–21: 18 regarding 14.01.2021).

When we consider that the indicator of the sixth SDG11 Target is the regular collection and 'adequate final discharge' of 'urban solid waste generated by cities', it is not difficult to see that at least some of the spatialised interactions of homeless people instantly render them collaborators of that Target. It does not matter that the concept of urban sustainability as such is absent from their verbal accounts about their daily work activities. When viewed from Goffmanian lenses, the situational nature of the patterned 'conducts' implicit in face-to-face interaction is more important than their conscious or unconscious performance (see, among others, Goffman, 1959: 29, 1963: 28). While our qualitative data cannot provide a precise measure of the extent to which homeless people contribute to UN Target 11.6, it does indicate that the environmentally sustainable interactions involved in waste collection by homeless people in the São Paulo public spaces could have played a role in figures such as the reported '700,000 tons of recyclable waste' collected by 'autonomous waste pickers' in the city in 2019 (Cseh et al., 2022: 128). In fact, waste pickers who independently sell the collected waste under dire economic conditions then accounted for 90% of all recyclable waste collected in São Paulo (Cseh et al., 2022: 128).

While the everyday spatiality of public spaces as daily workplaces for homeless people explicitly reveals one ongoing interactional contribution to SDG11, the two spatialities related to public spaces as everyday stages for homeless people's affectionate interactions with animals and plants offer less explicit clues. Almost half of our textual data (i.e., thirteen instances) contains verbal references to affectionate interactions featuring (seven) cisgender men, (two) cis and (two) transgender women, as well as (two) children alongside pets—primarily dogs (twelve times) and one cat—in front of our team. Squares and sidewalks appeared in the *corpus* (three times) as usual places for keeping dogs, always spotted close by their owner's side (Fig. 4.3).

The affective role of animals in the everyday life of homeless people becomes evident, among others, in the struggles Eva relayed to Patitucci. Describing her kitten as 'my little daughter', Eva preferred to stay in the streets with the cat rather than move to a city shelter for women and families without her (FD Patitucci 2020–21: 68 regarding 08.01.2021).

Our team also described four explicit (non-)verbal expressions of care towards dogs either in downtown squares (three times) or on a park sidewalk in the so-called expanded downtown area of São Paulo. Patitucci witnessed a poignant scene during a dusk in mid-November 2020 at the historically oldest city site, a square built around the remains of the foundational Jesuit school from 1554. Holding two lunch boxes in his hands, one homeless man approached Rita, the homeless woman Patitucci was speaking with in front of her tent. Rita opened the boxes and placed them on the ground, for the 4 dogs sleeping inside the tents to eat 'when they wake up' (FD Patitucci 2020–21: 22 regarding 18.11.2020).

One last facet of the interactional dynamics at play concerns explicit bodily expressions of affection towards animals (three times), which our team spotted in squares. During her fieldwork in the city Cathedral Square at noon on a weekday in that same November 2020, Anna Silva witnessed a homeless 'man who had just come out of the tent [and] went back to play with the dog tied to a leash. After filling a pink bowl with water for the dog to drink, the man continued on his way across the square' (FD Silva 2020–21: 11 regarding 16.11.2020).

An analogous spatiality stands out in our data set regarding the affectionate interactions particularly of homeless men with plants on sidewalks (three times). Still in November 2020, but on a weekday morning in the Mooca district—the downtown neighbourhood with the second-largest amount of homeless people apart from the historical centre (Qualitest, 2021a: 18)—, Santos noticed green plants on 'both sides of the [João Tobias] street. Some plants were in pots, and others in cans. The plants lent a cozier atmosphere to

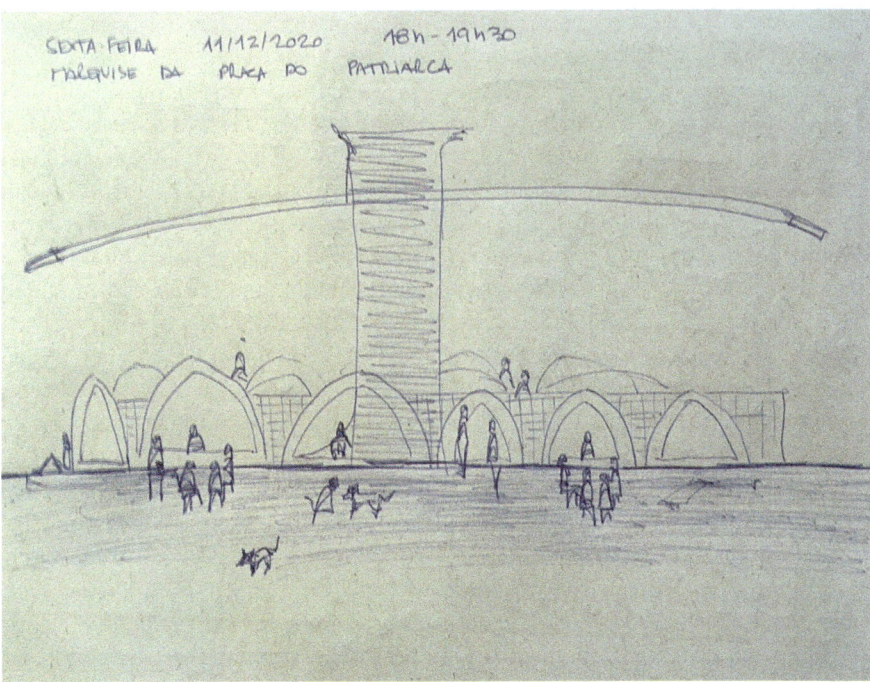

Fig. 4.3 Homeless people and dogs in front of a tent gathering—Patriarca Square, Friday 11.12.2020. (© Giulia Patitucci—SMUS Team)

the place, dispelling the greyness of the sidewalk and tarmac and adding to the colourful nature of the tents' (FD Silva 2020–21: 24 regarding 27.11.2020).

Almost two months later, the same Santos took note of the 'happiness' of the homeless man Martin when asked about a blackberry tree next to his tent: '"This is my little baby, the love of my life". He talked to the plants while showing me what was in the pots' (Fig. 4.4).

Apart from playing an effective role in Martin's everyday, his placing of plants 'in front and around the tent' accomplished an environmentally protective task. When asked to elaborate further, Martin said that 'it's because of the litter that many people throw'. Seeing the trees as an 'expression of care for that place', people would refrain from littering near his tent, where he began planting two years ago when he arrived (FD Santos 2020–21: 67–68 regarding 14.01.2021).

Although sparse, these ethnographic observations contribute to existing qualitative accounts regarding the interactional relevance of plants in the daily routine of São Paulo's homeless people (Frehse, 2021: parts 2 and 3). Furthermore, we must not forget the studies on the affective bonds of homeless people with animals (Bailey et al., 2023; Cleary et al., 2020).

For the purposes of this chapter, the interactional spatialisation of affection towards animals and plants in public spaces heightens a second, still unexplored dimension of environmental sustainability. In their daily affectionate interactions for caring for non-human beings such as pets and plants, homeless people inadvertently foster *environmentally inclusive* interactions. In addition to addressing the previously mentioned Target 11.6, which pertains to air quality and waste management, the affectionate interactions discussed here also pertain to Target 11.7, which focuses on the promotion of 'inclusive green and public spaces'. However, in this case, the Target recipients of inclusion are non-human environmental subjects rather than 'women and children, older persons and persons with disabilities' as stated in SDG11. Let us recall that the UN definition of 'social inclusion' refers to the 'process of improving the

Fig. 4.4 Martin's tent and plants—Street sidewalk in Mooca district, Thursday 21.01.2021. (©Valéria Jurado + Ednan Santos—SMUS Team)

terms of participation in society of people who are disadvantaged on the basis of age, sex, disability, race, ethnicity, origin, religion and economic or other status through enhanced opportunities, access to resources, voice and respect for rights' (UN, 2016: 20). In short, social inclusion encompasses the planetary-wide spectrum of (non-)human difference. Through their daily interactions with pets and plants in São Paulo's public spaces amidst their profound vulnerabilities, homeless people unmistakably demonstrate their tacit, albeit proactive, contribution to expanding the anthropological scope of social inclusion. They transform the environmental dimension of social inclusion in public spaces into an urban sustainability issue.

4.4.2 One Future Prospect of Socially Inclusive Public Spaces

While the ethnographic data set forged with the aid of the SMUS Toolkit during face-to-face interactions with homeless people (Phase 1) has enabled us to identify two *already ongoing* environmental contributions of homeless people to SDG11, the use of the SMUS 'glasses' during the aforementioned training course (Phase 3) brought to the interpretive forefront a *future* prospect of more inclusive public spaces. In this context, what holds significant analytical importance are the spatialisation patterns of interaction implicitly present in three of the aforementioned everyday spatialities of homelessness within public spaces. Our team duos communicated these issues both verbally and visually to the practitioners during their various face-to-face interactions throughout the course, which became spatial either on each of the four institutional sites or in the streets and squares. After being encouraged to make strange their own previous conceptions about homeless people while reflecting about the spatialities in focus, 8 of the 26 practitioners have emerged as particularly significant for the objectives of this paper. In our field diaries, we find 16 verbal observations made by three female and five male social workers about their newly acquired knowledge on these spatialities during those unusual pedagogic situations of ethnographic interaction.

The first spatiality addressed in the course was the social ordering of the public space as dwelling place. The ethnographic exchange regarding this issue over the following weeks prompted our team to compile six field notes documenting the practitioners' newly acquired knowledge. One of these excerpts reads as follows:

As we were walking past *Bar do Peixe*, [the social worker] Regiane [Garcia] recalled Juca's story, which she had shared with me the previous week: this homeless man has picked a fight with the bar and a Neo-Pentecostal church located across the sidewalk from the alley he used to dwell because they want him out of there. I then asked her: "What do you think that this space represents to Juca, to warrant such a fight? "It's his *home* [casa]", she replied in a eureka-like verbal outburst. (FD Reis 2021–22: 34 regarding 02.12.2021)

Two additional excerpts within this dataset underscore the idea of public spaces as being the dwelling places of homeless people. They refer to the newfound awareness of two other social workers, Patricia and Emerson, who came to realise that portable personal belongings carried by homeless people, such as 'tents' and 'what they carry on their backs', constitute their 'home'—as revealed by the metaphorical sense of used term '*casa*' in Portuguese (DaMatta [1987] 1995; see FD Quintão and Cunha 2021–22: 20 regarding 03.12.2021; FD Patitucci 2021–22: 52 about 21.12.2021). These accounts signal that the interactional dynamics facilitated by the SMUS Toolkit resulted in a discernible broadening of the practitioners' cognitive horizon concerning the daily bodily use that homeless people make of public spaces by means of their interactions with material objects.

Four other text excerpts of fieldnotes revolve around urban public spaces as places of either family-based interactions. In the corresponding course session, Regiane commented on family relationships from various angles. The strong impact of the conversation on her became evident the following day when Reis received the following captioned photograph, which he included in his field diary (Fig. 4.5).

Indeed, at the end of the aforementioned session, our team duo had invited Regiane and Patricia to

Fig. 4.5 '[T]his image reflects the ways in which a family is conceived in the street… Set up in tents as if it were a village, as if they were neighbours. Those by my side take care of me'—Republic Square on Thursday 16 December 2021. (© Regiane Garcia; FD Reis 2021–22: 90 regarding 16.12.2021)

'take photos' about the 'family relations of homeless people'. Only Regiane accomplished the task.

The verbal accounts of the practitioners who joined other team duos follow a similar cognitive path. According to three former homeless men who later became CBO social workers, in the streets a family takes shape irrespective of 'blood ties', forged by people 'living together'; hence 'I could choose my family' (see João Nery, Robson Mendonça and José Carvalho in FD Quintão and Cunha 2021–22: 36 regarding 26.12.2021).

This type of reasoning often coincided with comparisons to past or ongoing conflicts within their biological families. This comes as no surprise when we consider, on the one hand, that in October 2021, nearly 35% of São Paulo's homeless people cited 'family conflicts' as the primary reason for dwelling in the streets or city shelters (Qualitest, 2021b: 35). On the other hand, dwelling in the streets implies the daily (re-) establishment of bonds within 'street families', each with its associated identity markers (Frehse, 2013: 120, 2021a: 53).

Moreover, this reasoning was nearly always (in three out of four instances) intertwined with references to the third and final spatiality under consideration here: public spaces as places of affectionate interactions with pets. Both female social workers and the former homeless CBO coordinator acknowledged cats and dogs as family members. Patricia summarised it concisely: 'sometimes the pet is also a family' (FD Patitucci 2021–22: 52 regarding 21.12.2021; see also FD Patitucci 2021–22: 38 regarding 15.12.2021; FD Quintão and Cunha 2021–22: 35 regarding 16.12.2021). Indeed, animals are always close by. Robson and the social worker Emerson understood them as 'companions' of homeless people in the streets, whereas Patricia saw them as their 'protection' (see FD Quintão and Cunha 2021–22: 53 regarding 16.12.2021; FD Santos 2021–22: 28 regarding 08.12.2021; FD Patitucci 2021–22: 52 regarding 21.12.2021).

The three spatialities brought to the analytical forefront through the SMUS Toolkit reveal three knowledge contents implicit in the spatialisation patterns of homeless people's interactions in public spaces. By comprising dwellings, families and pets, these contents trigger the daily interactions of homeless people with an array of other material objects and living beings, which are not being addressed here.

Hence, this collaboratively developed spatial-methodological ethnographic knowledge carries other cognitive implications. As practitioners apply this knowledge in their day-to-day professional practice *with* and concerning homeless people, they become active agents of SDG11, above all Target 11.1, which revolves around initiatives towards 'adequate' housing with 'sufficient living area'. After all, the knowledge at stake here is socially inclusive by its very nature. If taken seriously in public policy and housing design, it simultaneously broadens the spectrum of (i) the social profiles of homeless people eligible for social housing and (ii) the challenges that must be addressed in the architectural program. Invariably, a sufficient living area must include designated spaces for personal backpacked belongings and handcarts, accommodating pets and catering to the needs of families formed through street partnerships.

It is precisely these facets that have led us to assert that the socially inclusive contribution of our practitioners to SDG11 lies in the future. To ensure a meaningful impact of this socially inclusive knowledge on a public-political level, it is imperative that the knowledge collaboratively forged by practitioners, like those we worked with in a transdisciplinary capacity, is made visible not only by public authorities but also by the academic community.

Our transdisciplinary project aims to be a spatial-methodological step in this direction.

4.5 Interactional Lessons Learned

Now that the methodological path pursued in the SMUS pilot project has been concluded also in analytical and interpretative terms, the lessons we learned along the way may be summarised as follows:

- *Regarding the SMUS pilot project:*
  1. The very conception and writing of this six-handed chapter introduce a fifth and

final phase to the SMUS project: one precisely of *analytical and interpretative crisscross* of *inter-transdisciplinary interactions*. This paper not only results from extensive transdisciplinary interactions involving a mid-career academic (Frehse), a junior academic (Reis), and a young practitioner with academic background (Patitucci), but it also integrates a wide array of more or less 'applied' social sciences such as sociology, anthropology, social work, architecture and urban design. Moreover, it resides at the intersection between academic teaching and learning, providing insights into the process of translating transdisciplinary research-practice data into academic writing.

- *Regarding the SMUS Toolkit:*
  2. Its practical-empirical implementation depends on
     (a) Developing communicational means that make it easier to present the purpose of the Toolkit to both practitioners and their respective institutional instances (public authorities, NGOs, CBOs) *prior* to the start of the transdisciplinary training. Hence, they are won for the cause and understand why they should get involved in the transdisciplinary training. The 'glasses' metaphor communicated especially well in this context, helping our winning the collaborating practitioners and their bosses for the cause of 'experiencing an alternative view of their daily work routine regarding homeless people'.
     (b) The situational establishment of a specific space-time *apart from* the practitioners' daily routine, so that the envisaged ethnographic exchange between researchers and practitioners about their everyday spatialities, and hence the transdisciplinary co-production of knowledge triggered by the SMUS Toolkit, may take place properly.
     (c) The researchers' ethnographic sensitivity to the variable suitability of specific tools within the twofold SMUS Toolkit techniques. Our collaborating practitioners showed greater resistance to experimenting with drawing and mapping techniques. Photography was the most well received means of non-verbal recording.
     (d) The continuous motivation and engagement of the practitioners based on the researchers ethnographic understanding of and solidarity towards the practitioners' expectations and frustrations regarding their own professional everyday life. For this purpose, Freire's problem-posing education method has proven crucial.

- *Regarding SDG11:*
  3. As we tackled these questions through the lens of the SMUS 'glasses', viewing the world from behind the shoulders of homeless individuals and the practitioners working with them in Covid-19-era São Paulo, the temporal challenges inherent in SDG11 came into clear focus. The very notion of an 'agenda' inherently implies the advancement of contemporary initiatives against homelessness to ensure a sustainable urban future. However, what if, during this interim period, homelessness not only endures but actually escalates in urban public spaces? To put it more accurately, how can we effectively address the daily reproduction of homelessness as a sociospatial phenomenon? Before the future unfolds, the everyday spatialities of homelessness, examined through a transdisciplinary approach using the SMUS Toolkit, can offer qualitative insights into two temporal contributions of homeless people to SDG11:
     (a) Homeless people are *already* contributing to SDG11 by means of environmentally sustainable and inclusive interactions in urban public spaces. Forged amidst their profoundly vul-

nerable everyday life, these contributions are as tacit as enduring. This is due to their inherently interactional nature.
   (b) Practitioners focused on homeless people *might* contribute to SDG11 if they were methodologically equipped with a spatial sensitivity rooted in the ethnographic perspective. Under these conditions they become agents for socially inclusive knowledge regarding the housing prospects of homeless people.
4. Our findings confront SDG11 stakeholders with the public-political challenges involved in the 'recognition' dimension of social justice (Honneth, 2004). After all, the social phenomena implicit in the three urban-sustainable contributions under consideration have a temporally and spatially transitory nature. How can we acknowledge the public-political significance of mere patterns of face-to-face interaction and the knowledge developed within their provisional spatial-situational boundaries?
5. Our findings emphasise that the discourse on urban sustainability encompasses an interactional-spatial dimension, so to speak. The most locally rooted, evidence-based contributions to SDG11 take place unconsciously, emerging within the transient and situational stream of face-to-face interactions.
6. Contemporary research on homelessness has emphasised the importance of an economically 'sustainable foundation' for a successful transition out of homelessness (Pleace, 2023: 237, 262). With the aid of our spatial-methodological toolkit, we invited both homeless people and practitioners to bring other dimensions of sustainability to the transdisciplinary forefront. This encompassed existing sustainable practices, whether involving ongoing patterns of environmentally sustainable and inclusive interactions or the co-creation of socially inclusive knowledge for the urban future.

The small quantitative scale of these findings is inherent to the very nature of ethnographic research. Precisely the SMUS Toolkit's capacity to embrace diverse perspectives favours its applicability in both inter- and transdisciplinary discussions on urban sustainability worldwide. While our collaboration began with homeless people and their assisting practitioners in the streets of São Paulo during the pandemic, the scope has since expanded to include other subjects (GCSMUS, 2023d). Hence, the SDG11 agenda as such turns into a spatial-ethnographically operational setting: a global set of face-to-face interactions *here and now*.

Acknowledgements We thank SMUS for funding the student-researchers-practitioners Ana Gil, Anna Silva, Ednan Santos, Giovanna Bernardino, Paula Quintão, Tales Cunha alongside Reis and Patitucci, who under Frehse's supervision conducted the ethnographic fieldwork underlying this chapter in the framework of the SMUS Action 4 coordinated by Frehse (https://gcsmus.org/action-speakers-for-action-4/). Our gratitude goes to each team member as well as to the homeless people and practitioners who joined the project; moreover, to Valéria Jurado for contributing with one fieldwork photo, and to Ignacio Castillo Ulloa, the scientific coordinator of SMUS Action 4. Partial funding for this paper was provided by the *Coordenação de Aperfeiçoamento de Pessoal de Nível Superior—Brasil (CAPES)—Finance Code 001. SMUS is funded* by the German Federal Ministry for Economic Cooperation and Development (BMZ) via the *German Academic Exchange Service* (DAAD)—Project Nr. 57526630.

Sources: Field Diaries

Bernardino, Giovanna. 2020–21. São Paulo: typed manuscript, 134 p.
Bernardino, Giovanna. 2021–22. São Paulo: typed manuscript, 41 p.
Cunha, Tales. 2020–21. São Paulo: typed manuscript, 99 p.
Cunha, Tales. 2021–22. São Paulo: typed manuscript, 26 p.
Gil, Ana. 2020–21. São Paulo: typed manuscript, 130 p.
Silva, Anna. 2020–21. São Paulo: typed manuscript, 80 p.
Patitucci, Giulia Pereira. 2020–21. São Paulo: typed manuscript, 99 p.
Patitucci, Giulia Pereira. 2021–22. São Paulo: typed manuscript, 55 p.
Quintão, Paula. 2020–21. São Paulo: typed manuscript, 67 p.
Quintão, Paula. 2021–22. São Paulo: typed manuscript, 78 p.

Reis, Caio. 2020–21. São Paulo: typed manuscript, 81 p.
Reis, Caio. 2021–22. São Paulo: typed manuscript, 129 p.
Santos, Ednan. 2020–21. São Paulo: typed manuscript, 87 p.
Bernardino and Santos. 2021–22. São Paulo: typed manuscript, 40 p.
Gil and Silva. 2021–22. São Paulo: typed manuscript, 77 p.
Quintão and Cunha. 2021–22. São Paulo: typed manuscript, 58 p.

References

Anderson, N. ([1923] 1967). *The hobo*. The University of Chicago Press
Bailey, C., Hockenhull, J., & Rooney, N. (2023). 'A part of me': The value of dogs to homeless owners and the implications for dog welfare. *Zoophilologica. Polish Journal of Animal Studies.*, special issue, 1–32.
Bartlett, L. (2008). Paulo Freire and peace education. In *Encyclopedia of peace education*. Teachers College, Columbia University. http://www.tc.edu/centers/epe/. Accessed 9 Aug 2023
Casey, L., & Stazen, L. (2021). Seeing homelessness through the sustainable development goals. *European Journal of Homelessness, 15*(3), 63–71.
Cleary, M., Visentin, D., Thapa, D., West, S., Raeburn, T., & Kornhaber, R. (2020). The homeless and their animal companions: An integrative review. *Administration and Policy in Mental Health and Mental Health Services Research, 47*, 47–59.
Cseh, A., Carvalho, I., Vallin, I., & Gonçalves-Dias, S. (2022). A coleta seletiva no município de São Paulo. In S. Gonçalves-Dias, L. Ziglio, & A. Cseh (Eds.), *Coleta seletiva de resíduos sólidos urbanos: Experiências internacionais e nacionais* (pp. 111–132). São Paulo.
DaMatta, R. ([1987] 1995). *A Casa & a Rua*. Rocco
Deckard, F., & Auyero, J. (2022). Poor people's survival strategies: Two decades of research in the Americas. *Annual Review of Sociology, 48*, 373–395.
Dias, S. (2016). Waste pickers and cities. *Environment & Urbanization, 28*(2), 375–390.
Dymitrow, M., & Ingelhag, K. (Eds.). (2019). *Anatomy of a 21st-century sustainability project*. Mistra Urban Futures & Chalmers University of Technology.
Fahlberg, A., Vicino, T., Fernandez, R., & Potiguara, V. (2020). Confronting chronic shocks: Social resilience in Rio de Janeiro's poor neighborhoods. *Cities, 99*.
Farha, L. (2019). *Guidelines for the implementation of the right to adequate housing (A/HRC/43/43)*. United Nations. https://digitallibrary.un.org/record/3872412. Accessed 16 Aug 2023
FEANTSA. (2023). *About us—What is FEANTSA*. https://www.feantsa.org/en/about-us/what-is-feantsa. Accessed 17 Aug 2023.
Frehse, F. (2006). Potencialidades de uma etnografia das ruas do passado. *Cadernos de Campo, 15*(14–15), 299–317.
Frehse, F. (2013). A rua no Brasil em questão (etnográfica). *Anuário Antropológico, 38*(2), 99–129.
Frehse, F. (2016). Da desigualdade social nos espaços públicos centrais brasileiros. *Sociologia & Antropologia, 6*(1), 129–158.
Frehse, F. (2018). On the everyday history of pedestrians' bodies in São Paulo's downtown amid Metropolization (1950–2000). In B. Freire-Medeiros & J. O'Donnell (Eds.), *Urban Latin America* (pp. 15–35). Routledge.
Frehse, F. (2020). Introduction. In F. Frehse, L. Kohara, C. Santana, & M. A. da Costa Vieira (Eds.), *Critical report—UrbanSus seminar: Dwelling in the São Paulo streets during the Covid-19 pandemic: Experiences, interventions, research*. IEA-USP. https://gcsmus.org/wp-content/uploads/2020-Frehse_Kohara_Santana_Vieira-Critical-Report-UrbanSus-I-ENG.pdf
Frehse, F. (2021a). The historicity of the refiguration of spaces under the scrutiny of the pre-Covid São Paulo homeless pedestrians. In A. Million, C. Haid, I. Castillo Ulloa, & N. Baur (Eds.), *Spatial transformations* (pp. 46–59). Routledge.
Frehse, F. (2021b). On the spatialities of the homeless' street (in Covid-19 São Paulo). In D. Talesnik & A. Lepik (Eds.), *Who's next? Homelessness, architecture, and cities* (pp. 92–97). ArchiTangle GmbH.
Frehse, F. (2022). Introduction. In F. Frehse, C. M. Reis, & I. Castillo Ulloa (Eds.), *Critical report—UrbanSus seminar: Everyday spatialities of dwelling in the streets of covid-19 São Paulo: Articulating research and practice*. IEA-USP. https://gcsmus.org/wp-content/uploads/2022-Frehse_Ulloa_Reis-Critical-Report-UrbanSus-II-ENG-1.pdf
Frehse, F., & director. (2021). *Street architecture in Covid-19 São Paulo*. Architekturmuseum der Technische Universität München. 23 min. https://www.youtube.com/watch?v=a-o3aaE22mM&ab_channel=ArchitekturmuseumderTUM
Freire, P. (1967). *Educação como Prática da Liberdade*. Paz e Terra.
GCSMUS (Global Center of Spatial Methods for Urban Sustainability). (2021). *SMUS action 4—exchange*. https://gcsmus.org/action-4-exchange/. Accessed 11 Aug 2023.
GCSMUS (Global Center of Spatial Methods for Urban Sustainability). (2022). *SMUS output/media: Action 4—Exchange*. https://gcsmus.org/output-media/actions/action-4/. Accessed 11 Aug 2023.
GCSMUS (Global Center of Spatial Methods for Urban Sustainability). (2023a). *SMUS Action 4—Exchange: Pilot Project 2020–2022: "Spatial Methods in Action: Everyday Spatialities of Homelessness for Urban Sustainability"*. https://gcsmus.org/action-4-exchange-2/. Accessed 11 Aug 2023.
GCSMUS (Global Center of Spatial Methods for Urban Sustainability). (2023b). *SMUS action 4—exchange: PEIPs 2022–2023*. https://gcsmus.org/peips-2022-202/. Accessed 11 Aug 2023.
GCSMUS (Global Center of Spatial Methods for Urban Sustainability). (2023c). *Pilot project (2020–2022): Everyday spatialities of homelessness*. https://gcs-

mus.org/output-media/everyday-spatialities-of-homelessnes/. Accessed 11 Aug 2023.
GCSMUS (Global Center of Spatial Methods for Urban Sustainability). (2023d). *PEIPs 2022–2023.* https://gcsmus.org/peips-2022-202/. Accessed 11 Aug 2023.
Genz, C., & Tschoeppe, A. (2021). Zur Erforschung von Räumen und Raumpraktiken. In A. J. Heinrich, S. Marguin, A. Million, & J. Stollmann (Eds.), *Handbuch qualitative und visuelle Methoden der Raumforschung* (pp. 225–236). UTB.
Goffman, E. (1959). *The presentation of self in everyday life.* Anchor.
Goffman, E. (1963). *Behavior in public places.* The Free Press.
Gregori, M. F. (2000). *Viração.* Companhia das Letras.
Gutberlet, J. (2021). Grassroots waste picker organizations addressing the UN sustainable development goals. *World Development, 138,* 105195.
Harvey, D. (2012). *Rebel cities.* Verso.
Honneth, A. (2004). Recognition and justice: Outline of a plural theory of justice. *Acta Sociologica, 47*(4), 351–364.
Kagan, S. (2019). Retracing my steps: A 10-YEAR journey to walking-based transdisciplinary research. *World Futures, 75*(4), 242–259.
Low, S. (2017). *Spatializing culture.* Routledge.
Low, S. (2023). *Why public space matters.* Oxford University Press.
Ministério (da Saúde). (2022). *Boletim Epidemiológico Especial Doença pelo Novo Coronavírus—Covid-19 N° 111—Boletim COE Coronavírus.* https://www.gov.br/saude/pt-br/centrais-de-conteudo/publicacoes/boletins/epidemiologicos/covid-19/2022/boletim-epidemiologico-no-111-boletim-coe-coronavirus/view. Accessed 9 Aug 2023.
Ministério (da Saúde). (2023). *Painel coronavírus.* https://covid.saude.gov.br/. Accessed 16 Aug 2023.
Morris, S. (2020). Disaster planning for homeless populations: Analysis and recommendations for communities. *Prehospital and Disaster Medicine, 35*(3), 322–325.
Park, R. E. ([1925] 1967). The city: Suggestions for the investigation of human behavior in the urban environment. In R. Park & E. Burgess (Eds.), *The City* (pp. 1–46). University of Chicago Press
Pleace, N. (2023). Complex needs and housing first. In J. Bretherton & N. Pleace (Eds.), *The Routledge handbook of homelessness.* Routledge.
Qualitest. (2021a). *Censo da População em Situação de Rua do Município de São Paulo—2021: Produto V—Relatório completo.* Secretaria Municipal de Assistência e Desenvolvimento Social de São Paulo.
Qualitest. (2021b). *Censo da População em Situação de Rua do Município de São Paulo—2021: Produto IX—Perfil Socioeconômico.* Secretaria Municipal de Assistência e Desenvolvimento Social de São Paulo.
Salcedo, J. (2019). *Homelessness & the SDGs.* PowerPoint presentation. UN-HABITAT Housing Unit.
Simone, A. M. (2019). *Improvised lives.* Cambridge & Medford: Polity.
Speak, S. (2013). 'Values' as a tool for conceptualising homelessness in the global south. *Habitat International, 38,* 143–149.
UN (United Nations Framework Convention on Climate Change). (1992). Convention on climate change. *Estudos Avançados, 6*(15), 161–192.
UN (United Nations General Assembly). (2022). *Resolution adopted by the general assembly on 16 December 2021 (A/RES/76/133).* UN.
UN (United Nations). (2023). *What is sustainable development.* https://www.un.org/sustainabledevelopment/blog/2023/08/what-is-sustainable-development-2/
Von Schönfeld, K., Roberti, A. C., Lopes, B., & da Conceição, G. (2023). (Re-)valuing and co-creating cultures of water: A transdisciplinary methodology for weaving a live tapestry of blue heritage. *International Journal of Heritage Studies (early access).* https://doi.org/10.1080/13527258.2023.2234349
WCED (World Commission on Environment and Development). (1987). *Report of the world commission on environment and development.* UN Secretary-General.
Wetzels, M. (2021). (Raum-)Fokussierte Ethnographie. In A. Heinrich, S. Marguin, A. Million, & J. Stollmann (Eds.), *Handbuch qualitative und visuelle Methoden der Raumforschung* (pp. 251–262). UTB.
Zscheischler, J., Brunsch, R., Rogga, S., & Scholz, R. (2022). Perceived risks and vulnerabilities of employing digitalization and digital data in agriculture—Socially robust orientations from a transdisciplinary process. *Journal of Cleaner Production, 358,* 132034.

Fraya Frehse is Professor of Sociology at the University of São Paulo, where she coordinates the Center for Studies and Research on the Sociology of Space and Time (NEPSESTE) and acts as a Lead Partner and Action Speaker of SMUS. She is an alumna of the Alexander von Humboldt Foundation and life member of Clare Hall College (University of Cambridge). Her research focuses mainly on urban theory; body, public space and urbanisation; social inequality/poverty and urban (public) space; homelessness; intersectionality and space; space and time in sociology; urban sustainability and public space; cultural heritage; urban visual culture; sociology of everyday knowledge.

Caio Moraes Reis is a SMUS-funded PhD candidate in Sociology at the Faculty of Philosophy, Languages and Literature, and Human Sciences of the University of São Paulo (FFLCH-USP), where he has made his MA in Political Science (2019) and his BA in Social Sciences (Sociology, Anthropology, and Political Science) (2016). He is a researcher of the USP Center for Studies and Research in Sociology of Space and Time (NEPSESTE) and of SMUS. His research focuses on homelessness, poverty, death, and urban sustainability.

Giulia Pereira Patitucci is an architect and urban designer trained at the Faculty of Architecture and Urbanism of the University of São Paulo (2017), where she made a MA on the issue of housing in the lives of homeless people (2022). Between 2018 and 2022 she worked with public policies for homeless population at the São Paulo Municipal Secretariat for Human Rights and Citizenship. In 2023, she continued addressing the same issue as coordinator of the Executive Secretariat for Strategic Projects at the São Paulo City Hall before moving to Brasília to become project manager of the Programme for the Democratisation of Union Properties at the Ministry of Management and Innovation in Public Services from the Federal Government of Brazil in early 2024. In 2017, she worked at the Social Rent Program of the Sao Paulo Metropolitan Housing Company.

Open Access This chapter is licensed under the terms of the Creative Commons Attribution 4.0 International License (http://creativecommons.org/licenses/by/4.0/), which permits use, sharing, adaptation, distribution and reproduction in any medium or format, as long as you give appropriate credit to the original author(s) and the source, provide a link to the Creative Commons license and indicate if changes were made.

The images or other third party material in this chapter are included in the chapter's Creative Commons license, unless indicated otherwise in a credit line to the material. If material is not included in the chapter's Creative Commons license and your intended use is not permitted by statutory regulation or exceeds the permitted use, you will need to obtain permission directly from the copyright holder.

Ethno-graphy on the East Kolkata Wetlands: A Transformative, Transdisciplinary Tool in Protecting Urban Ecological Heritage

Jenia Mukherjee, Shreyashi Bhattacharya, and Sasidulal Ghosh

5.1 Introduction

We are in the Urbanocene (Palme & Salvati, 2021) with dramatic alterations in the rate, scale and shifting geographies of urbanisation (West, 2017; Myint, 2018). For a long time in academic scholarship, cities and nature had shared an antithetical relation to each other—the 'urban' meaning built infrastructures such as buildings, bridges, concrete homes and the 'environment' implying agrarian lands, pastures and water bodies. It is only during recent times that critical urban studies such as environmental history, political ecology, etc. have interrogated this Cartesian divide by arguing that cities are largely derived from their wider ecological surroundings (Schott, 2004; Mukherjee & Véron, 2023). This understanding has been foregrounded in SDG11.4, which aims at strengthening efforts to protect and safeguard the world's cultural and natural heritage.

This chapter describes the Global Center of Spatial Methods for Urban Sustainability (SMUS) sponsored Action 4 (Exchange) initiative, the Practical Empirical Implementation Project (PEIP) on the East Kolkata Wetlands (EKW) to advance SDG11.4 (in particular) by harnessing academia-practitioner exchange and engagement in co-designing a roadmap towards urban sustainability, through meaningful understandings and internalisation of this dynamic socio-ecological space—the urban ecological heritagescape of Kolkata. The SMUS 'glasses' (see Chap. 4, p. 125) deployed in the project are the metaphoric representation of the SMUS Toolkit, which we made use of in a creative way. Our SMUS glasses consist of two aspects: (1) ethnographic observation of the everyday spatialities of the researched subjects through participant observation, key informant interviews and focus group discussions that 'make familiar what is strange and strange what is familiar' (see Chap. 4, p. 121), and (2) visualisation techniques of these everyday spatialities through drawing, photograph, sketches or usage of technical devices that make it possible to visualise the everyday spatialities of the research subjects. The 'everyday spatialities' refer to the everyday determinations of the researched subjects, and their

J. Mukherjee (✉)
Department of Humanities and Social Sciences + SMUS, Indian Institute of Technology Kharagpur, Kharagpur, India

S. Bhattacharya
Rekhi Centre of Excellence for the Science of Happiness, Indian Institute of Technology Kharagpur, Kharagpur, India

S. Ghosh
State Fisheries Development Corporation Limited – West Bengal & Leaseholder of Jhagrashisha Bheri – East Kolkata Wetlands, Kolkata, India

© The Author(s) 2025
F. Frehse et al. (eds.), *Spatial Methods in Transdisciplinarity for Urban Sustainability*, Sustainable Development Goals Series, https://doi.org/10.1007/978-3-031-84367-9_5

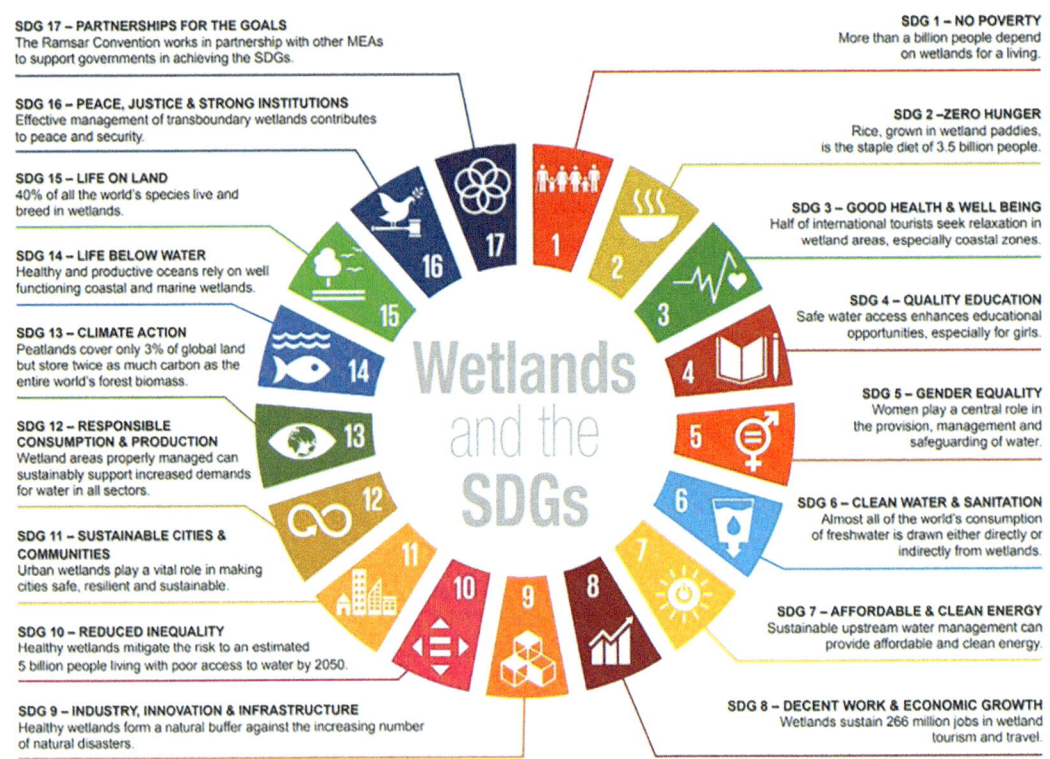

Fig. 5.1 Wetlands and SDGs. (Source: Ramsar Convention on Wetlands, 2018: 6

dynamic tangible and intangible interactions with the space (and the ecosystem) in which they live.

The EKW earned recognition as the Ramsar site.[1] This means that it is the world's largest recycling ecosystem, which treats 750 million litres of Kolkata's effluent each day through wise-use recovery practices involving indigenous communities and generating nature-based solutions (Ghosh, 2005; Kundu et al., 2008; Mukherjee, 2020). According to the Ramsar Convention guidelines of 2018, the protection and sustainable use of the EKW is crucial for helping India to realise its SDGs (Ramsar Convention on Wetlands, 2018). The Ramsar Convention Report of that same year (Ramsar Convention on Wetlands, 2018) highlighted interlinkages between wetlands and the SDGs, emphasising the need to preserve these crucial ecosystems at all levels (Fig. 5.1).

With its thrust on safeguarding the world's natural-cultural heritage, SDG11.4 highlights the need to trace the history of the co-evolution of urban sites and ecological infrastructures. Their 'sustainable flows' (Mukherjee, 2015a, b) ensure and enhance socio-ecological resilience (Bhattacharyya, 2018; Mukherjee & Bhattacharya, 2023b).

The PEIP team in Kolkata applied the SMUS glasses by combining ethnography/qualitative interviews with visualisation techniques. Here, we conceive that application as an *ethno-graphic* exercise Mukherjee & Bhattacharya, 2023a), given that it implies the integration of conventional ethnography with visual methods and formats—the hyphen sig-

---

[1] Ramsar sites are wetlands of international importance, designated by the Ramsar Convention, an international environmental treaty signed on 2 February 1971 in Ramsar, Iran, under the auspices of UNESCO. It is intergovernmental treaty that provides the framework for national action and international cooperation for the conservation and wise use of wetlands and their resources (Ramsar Designation, EKWMA; https://ekwma.in/ek/about-us/ramsar-designation/)

nifying an integrated demarcation rather than segregation of the two impactful methodologies. The aim of this method is to blur the rigid boundaries existing across academic disciplines that limit the inclusion of stakeholders and to incorporate collaborative insights and practices facilitating sustainable co-management of complex natural-cultural heritage spaces. Ethno-graphy integrates conventional ethnographic techniques such as participant observation, transect walks, interviews and discussions with visualisation techniques such as photography, videography, sketches, etc., to engage, immerse, interpret and present the multiple realities of a space (Mukherjee & Bhattacharya, 2023a). This method aims to enhance and enrich the conventional ethnographic approach, standing as a foundation for a new transdisciplinary methodology. After all, it gives agency to (non)human actors, acting as a participatory, agency-induced ethno-visual tool. So far as our particular project is concerned, the exercise was an enabler in fusing the historical context of the EKW with the spatial methods. Findings from transect walks, key informant interviews, focus group discussions and participant observations could be (re)interpreted through the application of visual tools such as photography, videography, illustrations, drawings, sketches, etc.

Through the spatial methodology of 'ethno-graphy', the everyday spatialities of the EKW and its inhabitants could be encapsulated and depicted, offsetting many familiar constructs and declensionist discourses dotting Kolkata's wastewaterscape that describe its ecological ingenuity being threatened by the urban expansion, overshadowing the intricate web of connections and collaborations existing between the city and nature. Projected as a contested space, due to the history of struggles and conflicts between the state, the city and the wetland inhabitants (Dembowski, 2000), the EKW awaited effective and impactful co-exploration at various levels, involving multiple stakeholders to unravel this (un)familiar urban ecological palimpsest. We consider our PEIP to be a crucial intervention in advancing SDG11.4, which focuses on preservation of nature-culture heritage by addressing their complex intersections.

The chapter builds on and describes an eight-month process (between February and September 2023), of a two step-methodology of the deployment of the SMUS Toolkit in the form of the ethno-graphic exercise in capturing and conveying chequered complexities surrounding Kolkata's peri-urban wetlands, facilitating, fostering and forging academia-student-practitioner exchange and engagement. In the first phase 16 students, enrolled in the Bamanghata and Kheadaha government schools located at the heart of the wetlands, immersed in co-learning activities and assignments across extensive application of the toolkit through various formats, determined by their rationale in the selection and combination of multiple spatial methods with the academic team. The students took the ethno-graphic toolkit to selected *bheris* (sewage-fed fish ponds), a distinguishable feature of the EKW and co-explored the space through deployment of the toolkit. The second phase requiring critical analysis of the data generated from the toolkit, through reflections and assignments, elicited nuanced understandings of the space from the students. The third phase focused on convincing the local practitioners about the efficacy of the SMUS glasses in multiple ways of knowing, encountering and becoming familiarised with the wetlands through transect walks, interviews, participant observations and group discussions, leading them to the last phase where these changes that have been cultivated would be evaluated by all the team members to assess the efficacy of the project. This chapter describes and explains this project process trajectory, collating and compiling exposure to and experience of 'ethno-graphy' by the PEIP team, the outputs-outcomes-impacts generated and envisioned, and the way forward in advancing SDG11.4.

5.2 The East Kolkata Wetlands

The East Kolkata Wetlands is a multifunctional wetland ecosystem encompassing 12,500 hectares of land, adjacent to the eastern side of

Kolkata, bordering the Salt Lake and the Rajarhat townships, that comprises almost 254 sewage-fed fisheries, agricultural land and solid waste farmland (Fig. 5.2). The EKW, also known as the peri-urban interface of Kolkata, is a Ramsar-recognised site for being the largest wastewater-fed aquaculture system in the world. It recycles urban effluent for free, adapts the role of a carbon sequestration sink and facilitates the growth of flora and fauna, while also generating a variety of provisioning services impacting livelihoods in the wetlands as well as the city.

The EKW is part of the vast network of tidal swamps in Kolkata, which emerged as part of British colonial attempts at hydraulic engineering and land reclamation in the Sundarbans delta—from where the city was reclaimed (Bhattacharyya, 2018). The city's effluent was initially channelled into the Bidyadhari River, which bordered the eastern margins of the city. However, due to an escalating sedimentation load, the river dried up and the system became defunct. The tidal-fed saline nature of the wetlands changed in 1939, when B.N Dey, then Chief Engineer of the Calcutta Corporation, introduced an elaborate internal drainage scheme, where the sewage of the city would travel via two channels—the Dry Weather Flow (DWF) and the Storm Water Flow (SWF), to their outfall Kulti River channel (replacing Bidyadhari), intersecting the wetlands through an intricate design of canals, inlets, sluices, sedimentation tanks, etc. (Mukherjee, 2020). The introduction of the sewage water gradually changed the aquatic environment of the region, transforming them from saltwater marshes to sewage-fed fisheries bereft of salinity.

The EKW presently treats around 750 million litres of wastewater daily, accounting for almost 80% of Kolkata's discharge, which is further used in the pisciculture and agriculture in the region, thus providing a unique cyclical relationship of interdependency between the city and the EKW (Mukherjee, 2015b). Although designated as an 'wetland of international importance' under the Ramsar Convention in 2002, the EKW now

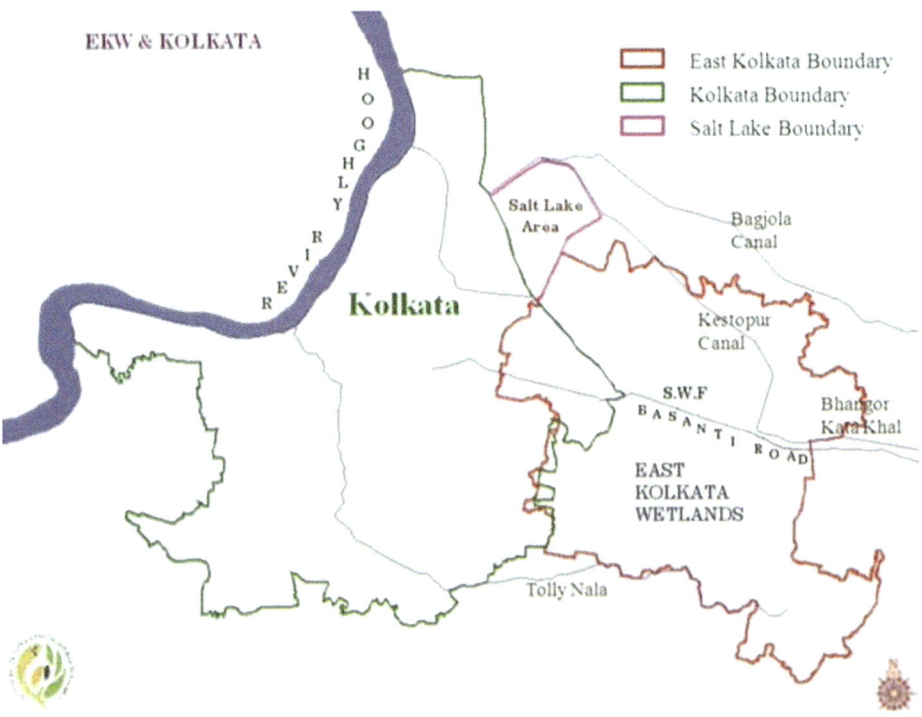

Fig. 5.2 Kolkata and the EKW. (Source: Mukherjee (2020)

faces the risk of illegal land conversion, which has unfortunately faded the memory of the once historical collaboration between city and nature and broadened the schism between the city and the wetland dwellers, with the latter considering the former as the principal antagonist.

5.3 Project Design and Team Composition

Trained in political ecology, environmental history and environmental humanities, the academic team from IIT Kharagpur and the architect from civil society, or 'creative collective' Disappearing Dialogues (DD), brainstormed and identified, listed and classified historical, qualitative and urban design methods and perspectives from archival and field analysis of the region with implementation scope in the project, capturing technical and ground realities through artistic dissemination, either as independent or integrative formats. The key agenda through the project design was to record and encapsulate spatial empirics towards better understanding of long-term measures in safeguarding and protecting the EKW, in collaboration and with involvement of local practitioners, in turn enhancing the social competence of the latter (Table 5.1). But how could this academia-practitioner exchange be forged and facilitate optimised impact—leading to effective outcomes beyond the life-cycle of the project?

We closely worked with 16 students from standard VIII from 2 government schools: Bamanghata and Kheadaha. These are the only schools administered by the state government of West Bengal, India, in the EKW that (i) are situated in the heart of the wetlands, (ii) 'wore' the SMUS glasses to explore the details of the EKW and (iii) made the local practitioners aware about the efficacy of the lens (Fig. 5.3).

The two fundamental reasons behind this team composition were the experience and expertise of DD to work with the students of wetland schools with an already established rapport with the school administration, and the conviction of academia about the active cognitive domain and cultural identity of the young adults as both knowledge absorbers and transmitters, which facilitated the process of knowledge co-creation and mobilisation in the EKW. The school students, nine males and seven females between 13–14 years of age, live in the vicinity of the schools and belong to marginal tribal communities, coming from households engaged in various EKW activities and urban labour for livelihoods, making the conversations and communications between students and practitioners much smoother.

The academia-practitioner collaboration with the young adults as 'ethno-graphers' (Illustration 3) was thus envisioned along a three-fold project execution plan:

(a) Design and development of an appropriate and feasible training-learning template by the Indian Institute of Technology Kharagpur and DD, for and with the 16 students.
(b) Deployment of the SMUS glasses in the 'ethno-graphic' format by the wetland school students in capturing the EKW as a dynamic and complex socio-ecological system through an interactive, immersive and integrative process.
(c) Co-exploration of the EKW using the toolkit—with local practitioners exposed to and convinced by the efficacy of the 'ethno-graphic' exercise; the knowledge products being finalised after detailed feedback from practitioners.

Table 5.1 Classification of the *bheri*/sewage-fed ponds sites of the PEIP

Name of the *bheri*	Ownership pattern	Personnel
Nalban Fisheries Project	Government/State owned	Soumen Chakraborty Project-in-Charge
Baro Chainavi Matsyajibi Samabaya Samiti	Cooperative	Gobinda Sardar Secretary
Jhagrashisha Bheri	Private	Sasidulal Ghosh Former Deputy Director, Department of Fisheries, Government of West Bengal Private leaseholder

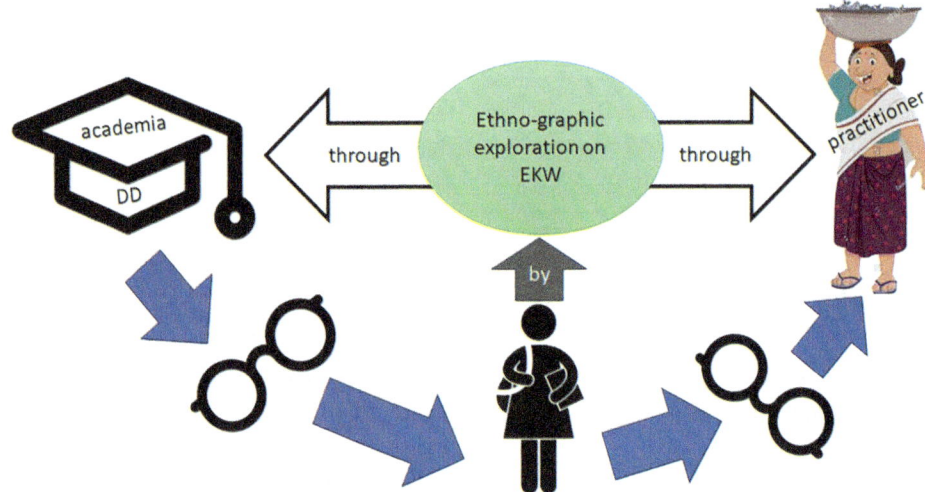

Fig. 5.3 Team composition—PEIP on the EKW. (Source: Authors)

Smart mobile phones were distributed to the students as 'ethno-graphic' gadgets performing multiple functions such as photography, videography, GIS tagging, audio-recordings of interviews, etc., in an easy-to-use, practical-useful manner. As DD had previously worked with the students on various projects, they had an idea about the interests, enthusiasm and expertise of the students. The first phase of selection was led by DD. In the next phase, the academic team oriented the students on the project objectives through an interactive classroom session on the EKW, after which the students were given homework to present their new learnings through an innovative interpretation. The resulting outputs from the students included essays, drawings and illustrations—which the academic team went through. From these outputs, 16 students were selected and finalised for the project. The academic team also pondered over the critical issue of whether it would be prudent to provide mobile Internet connection and access to the students, considering its dangers for tender adolescent minds. The two schools, situated in the heart of the wetlands, provide education to a diverse set of students (both male and female) from marginalised communities engaged in low-income activities, living in areas where Internet connectivity is sparse and where children do not have the privilege to afford mobile phones. So, providing them smartphones presented a great challenge for us, as we had to orient the young adults to use the device for constructive and educative purposes. When the need emerged as vital, especially for two specific reasons—interconnection with PEIP students and team members for conducting group activities, and quick access to digital data and images to work on the project, adding and verifying facts—the phones were enabled with SIM cards offering talk time and Internet data.

Technologically equipped and mentally charged up, the academia-DD-student teams embarked on the PEIP journey of venturing into the EKW for knowing and conveying her in an exciting and meaningful way.

5.4 Phase-Wise Project Execution

5.4.1 Transecting (Un)familiar Trails: 'Ethno-graphic' Training, Exchange and Exposure

Kolkata harms the wetlands…

The above remark stems from one of the students. But this was the most evident perception among the students regarding the city–nature relationship in the region—when we met them first. To them, the wetland was an ecological resource and the city was the enemy. It was neither uncommon nor unexpected, considering some direct experiences and the media coverage on the rising threats of the city's unchecked urbanity engulfing the natural boundary of the wetlands. This, added with recent trends on the declensionist portrayal of statecraft narratives along with preservation activism (Mukherjee, 2020; Mukherjee et al., 2022), presents a linear perspective on the city-nature relationship, with the city being portrayed as a ruthless and unforgiving aggressor (Mukherjee, 2020). However, despite these declarations, when probed deeper in the very first introductory meeting and exchange at the Bamanghata school premise, the students showed very little awareness on the actuality of the functioning of the wetlands co-habited by them. Borrowing from oral narratives, they quoted the benefits of wetlands as a pollution filter and an essential source of fish production for the city but had limited knowledge on how the space was created, its history and what is it surviving upon.

We wanted them to question the familiarity they had always been comfortable with, to rethink what they have always known, without being overwhelmed by the academic team's perception. So, we began orienting them to the project by letting them be curious, nudging them towards exploring their own history and gently probing them to ask questions about the past. We discussed with them the backbone of wetland economy, the *bheri* or sewage-fed fish ponds, that evolved from the convergence of colonial hydrology and local indigenous practices in 1939, when the city's wastewater was directed into the wetlands and used for pisciculture in the region. While the city got a free and natural sewage treatment plant, the wetlands of Kolkata became a source of livelihood and cultural practices. This new information befuddled the students, who at first couldn't believe that despite living beside the wastewater canals they weren't aware of its use and impact! This new information came as a shock to them, and we wondered if this could affect their confidence in the long term as the academic team, the outsiders. In their eyes, suddenly showcased more knowledge about their region than them, the inhabitants.

Conducting initial rounds of training and classroom sessions, the academic team exposed them to various qualitative methods but in a playful and storytelling narrative, where two of the academic team members (Shreyashi and Raktima, research scholars at IIT Kharagpur) presented their own qualitative research case studies to them. The focus of these sessions was to discuss and describe the method of conventional ethnography to the students, an essential component of the SMUS Toolkit, and using it as a guiding tool towards ethno-graphy. Shreyashi told the tale of the historic river Adi Ganga in Kolkata, using snippets from mythological accounts to colonial reports, presenting the need to study the history of a place, through archives and secondary literature, in order to understand its present condition. She showed them pictures of the river, taken during her transect walks, of her interviews and focus group discussions, and how she and her colleague, Debika, developed sketches on the river based on their participant observations depicting the variety of life on the riverscape. This presentation helped the students become acclimatised to various ethnographic methods such as transect walks, key informant interviews, focus group discussions and, seeing them developed into sketches as complementary research outputs, helped them grasp the concept of visualisation tools as it was something they would immediately connect with (Fig. 5.4), making it their first exposure to ethno-graphy:

Raktima's work in the dry fish sector in the Sagar Islands of the Indian Sundarbans showcased her journey in the field through photographs. As she relayed her trip to the field, dotted with interesting conversations and adventures, the major conjectures captured in photographs, such as the journey in the boat, fishing gear and women in work, were displayed in the background, interlinking the story with visual appendages (Fig. 5.5):

Fig. 5.4 Sketch shown during the Adi Ganga presentation. (Source: Bhattacharya et al., 2024: 8)

Fig. 5.5 Photo shown during the dried fish presentation. (Source: Bandopadhyay et al., 2022: 8)

Raktima's narrative also described the importance of participant observation,[2] a crucial aspect of ethnographic method that the students would have to internalise and apply during field visits, as well as in their own private time in their wetland community. Raktima explained how simply

---

[2] This is a qualitative methodology technique used by social scientists, mainly anthropologists and sociologists, and implies that the researcher completely immerses themselves in the day-to-day activities of the participants in the field, while observing, recording and analysing the collected data.

observing the scene and the participants, with her everyday presence in the scene, contributed to her gaining the trust of key informants and creating a baseline for questions before venturing into interviews and focus group discussions comprising formal–informal interactions with community groups to help her learn about individual as well as collective perceptions. The students, inhabitants of wetlands, could use these techniques in their home and community settings to maximise the potential of learning about the everyday realities of the space.

Subsequent sessions focusing on the technicalities of photographic and videographic interpretations of complex urban spaces through the mode of more creative and engaging pedagogies were conducted at the school premises by IIT Kharagpur and DD, showcasing the potential of visual interpretations in disseminating methodological outputs. The students were provided with art kits and exposed to different functions of smartmobile phones to practice and implement the conceptual discussions on ethnography. The discussions facilitated engaging interactions between the instructors from the academic team and the students who, considering their young age, felt overwhelmed by the various techniques, although they continued, very commendably, to ask the very basic but pertinent questions: *How and where would they apply these methods, and will this process contribute in safeguarding the wetlands?*

While the classroom sessions provided an in-depth understanding of research methods, they elicited varying reactions and responses from the students. The school students were presented with a new purview of their own space by the academic team, who were actually 'outsiders', and it required corroboration by people who they already knew and trusted through community kinship. Why would they want to learn about these methods? They are not seasoned researchers, but young inhabitants living on the periphery of a complex urban ecoscape, dotted with its own challenges. The presentations had intrigued them and showed them the importance of innovative tools in reaching out to audiences, but they needed to substantiate that information as well.

With that thought in mind the students were taken to different types of *bheris* to interact and cross-validate the information on the wetlands, as well as to test the efficacy of these new methods that they had learnt.

The students, guided by mentors from the academic team and DD, visited a government, a cooperative and a private *bheri*, or sewage-fed fisheries that are the lifelines of the wetlands and crucial in understanding the working mechanisms of the effluent-based livelihoods in the region, to apply the toolkit, where they interacted with practitioners through focus group discussions and interviews, mapped the space through transect walks and documented the manual-infrastructural dynamics through photographs and videos. The first visit to the field was to the cooperative *bheri*—the fishery Baro Chainavi Matsyajibi Samabaya Samiti, where students talked to the *bheri* secretary, Gobinda Sardar, who after understanding the novelty of the project engaged in a long discussion with the students about the wetlands history, narrating to them the minute details about the role of the city's sewage in the survival of pisciculture in the region, which is the backbone of wetland economy and the livelihood of the people involved in it.

Being an inhabitant of the region, Gobinda Sardar's narrative provided overwhelming details of local attachment to the space, phrases and emotions that the students could connect with, while learning about new information that was previously unknown to them.

The conversation, edging between factual and emotional narratives, helped the students gain confidence and probe deeper into the socio-ecological practices at the *bheri,* during the training, field visits and then later as well, showing that they had immersed themselves into the themes they had learned in the classroom. They would start conversations, take pictures and videos of the space without prompt from the experts and start categorising, which methods to use and when, based on their training and field experiences.

In the next few days, the students started to form groups to brainstorm about these methods, which was an assignment provided by the academic team to evaluate their understanding of the methodology. This allowed the school students to assess the capacity and effectiveness of the toolkit based on the field visit and by learning more about them on the Internet.

Interestingly, the students and the academic team would come up with themes to discuss, start classifying them into social, ecological and economic sections, and prepare questionnaires for their next visit. This took place in a state-owned *bheri*, the Nalban State Fisheries Development Corporation, the largest *bheri* in the East Kolkata Wetlands. During this visit the students began their field journey by applying the transect walk method, in groups, to explore the vast space of the *bheri*, with each member of the team assigned different tasks such as sketching a route of the area, taking photographs of the different people, infrastructures and apparatuses, preparing questions on what they had seen to discuss with *bheri* officials, etc.

Each team worked in parallel, sometimes diverging due to a difference in themes and academic team availability, and, guided by the mentors from the academic team and DD, converged and interacted with the workers in the field to collect information through informal discussions. The students noted down the questions that they would later validate with the Nalban *bheri* supervisor Soumen Chakraborty, who provided interesting insight on the wastewater mechanism of the wetlands, the role of water hyacinths in filtering the sewage, livelihood provisions and maintaining the flood balance of the region.

While it was a new experience for the students, who belonged to marginalised communities, to interact directly with state officials, it didn't deter their positionality as young ethnographers, leading them to carry on the conversation with their previously prepared questions, as well as the new ones that they generated on the spot using their presence of mind and based on their field exposures and explorations. The interview and group discussions allowed them to learn more about the mechanisms of running a *bheri*, an essential component in the survival of the region that the students only knew by name and passed conversations. Now they learnt about employee working hours, wages and challenges faced by a government *bheri*, such as inadequate supply of wastewater, and also about opportunities for further income generation through renting of the space for recreational activities and festivities. These interactions and exposures were facilitated, supervised and followed up by the academic team to ensure that the students were on track with the project. The iterative rounds of discussions, from classrooms to the fields to post-field communication, gave the students the opportunity to constantly reflect on their observations and analyse them accordingly, getting refined, critical and more nuanced with every approach.

5.4.2 Selections and Combinations of Methods with Narratives Taken as Agents of Change

The field activities helped the students immerse in the space as they investigated the nitty-gritties of the wetland dynamics as ethno-graphers. The next part of the project now required the critical analysis of the data, and so the challenge was no longer the deployment of the tools but the diagnosis of their applicability and implementation in data analysis. We wanted them to provide us reports on the collected data from their field understandings through the application of these methods, detailing the use of the tools in obtaining it, while also not trying to assert control over their narratives and creativity. So, we assigned them the task of explaining what they had seen and learned from the space in a way that was most comfortable for them, whether through a written report, a picture or a photo essay—whatever manner would showcase their findings in the best possible way.

The students took a few weeks to complete the assignment. The extended timeline was necessary to accommodate flexibility in balancing project deliverables and outputs with regular school work amidst an extreme heatwave affecting the city in the hot tropical summer. The students were enthused about the work but required some more time, so we let them have it, while checking in on them and their progress through WhatsApp.

In the meantime, the academic team prepared and scheduled the next sets of events, gently communicating the plans with the students via WhatsApp, answering their queries and encouraging them to innovate, yet without saturating their workload. However, the long time in between was affecting us all—the academic team apprehensive about losing gained momentum, and the students being out on their own, feeling a bit disoriented, which is very normal for young adults not conventionally trained in research and enduring the pressures of a school curriculum. After a couple of weeks, the students thankfully communicated about developing something that they would like to share with the academic team. This was an interesting moment for the team, as for the first time the students took the initiative to organise a workshop, which was held in the last week of May 2023.

This workshop was organised at a shared community space, a temple premise in the wetlands that the civil society members had arranged for the PEIP team to gather make a stocktake of the assignments and project progress. The temple space was chosen because of its space and cultural importance, as it is a centre of engagement and discussion in Hindu society, where the community elders and women would usually gather for chitchats and their daily reflections. We wanted to use this opportunity to engage the community with the school student's work. Community members ranging from elders, women and livelihood practitioners, such as fishery workers and farmers, were invited to participate in the discussion surrounding the project so that they could witness their children's work and provide practical feedback.

The programme began with the students coming forward in the groups they were previously assigned to and explaining their assignments. The three groups recapitulated their journey thus far, with their own interjections on what they have learnt and how impactful it had been. The first group narrated their changing perception of the space, which they always considered a segregated patch of ecology that required conservation, but their journey to date had exposed them to the wastewater mechanisms, the social dynamics and the religious attachments in the area, citing examples of the fish harvest festivals celebrated in the *bheri*. They further showcased videos of interviews with the community and small-scale *bheri* workers in their locality, which they had conducted on their own to understand more about the EKW.

The next group demonstrated their analysis using photographs and a sketch of the wetland *bheri*, trying to simulate the previous transect walk in front of the local community. They discussed how they engaged with the application of exciting techniques such as participant observation, interviews, etc. These facilitated their understanding of the different livelihood practices in the region, from the varieties of pisciculture and agriculture to introduction to new types of flora and fauna, and the role of local knowledge that has been passed on through generations ensuring survival and sustenance in the EKW.

"Fish cultivation requires a lot of hard work.
What we consider as nutritious, is their livelihood!"

This remark is a representation of the collective understanding of the students of the pisciculture scenario in the EKW. Many students echoed this sentiment. They started by recognising the interconnectedness between ecology and social-economy in the region, as noticed in the above remark, that brought to the fore their changing outlook on their habitat—seeing it beyond the popular discourse of a protected space, and as a vibrant and bustling ecosystem, sheltering multiple actors, agencies and enactments.

This understanding was taken further by the last group who explained their understandings of the methods and their application through a prototype of the Nalban *bheri* (Fig. 5.6), developed by them using household and stationery items from the art kit:

The students asserted that when they attempted to develop a report to describe their Nalban experience, they didn't feel much excitement. As they looked back at the photographs and videos of the day in their homes, they felt the need to impart similar visual experience to the

Fig. 5.6 3D model representation of the Nalban *bheri*. (Source: PEIP student team)

audience, to make them relate with what they had seen and understood. With this urge and motivation, the students decided to develop a 3D cardboard model of the Nalban *bheri*, that represented a miniature understanding of the space. It showcased the inlet and outlet canals, the fishing ponds, tools and equipment, hatchery tanks and cold storage units with a silhouette of the city skyline in the backdrop, all of which was produced by recycling leftover cardboard, styrofoam and household materials, such as cotton, straw and rice grains, and art tools provided by the academic team to depict the operation of the state-owned fishery. The students realised that a 3D model would help in actuating the vastness and scalability of the space that can help disseminate this knowledge to an unknown audience. This model represented the student's recollection of the space, mapping the region, its various infrastructures and their usage, and the linkage with the city, which they explained in a lucid and comprehensive manner, fulfilling all the criteria of the ethno-graphy approach. This miniature model captured the very essence of the project—making the familiar strange and the strange familiar—where an integral part of the EKW ecosystem that was less known by the inhabitants was afforded clarity in their minds through the use of the spatial methods applied by them.

The final part of the day on the temple premise comprised local community members' reactions and feedback to the students' presentations, an impromptu reaction that was followed by their own expositions, enlightening the PEIP team. The women conveyed their experiences of working in the *bheris*, which is otherwise considered a male-dominated space. They also unravelled how *bheris* are more than just fisheries and also brought out utility challenges for the wetland inhabitants. These critical reflections were helpful in assessing project progress and reorienting the project activities, which would now require a more in-depth understanding and internalisation of the wastewaterscape, its challenges and opportunities, using the SMUS glasses, in turn, to be carried forward and taken to the local practitioners at the next level.

5.4.3 The Trained as Trainers: Academia-School-Practitioner Engagement Through 'Ethno-graphy'

The third phase and most critical juncture of the project was when the young inhabitant ethnographers, now taking reign of the SMUS glasses, started working with and instructing practitioners about its application. The stage was set on 27

August 2023, when the PEIP team organised the academia-school-practitioner engagement workshop in an auditorium of eastern Kolkata, where the school students shared their findings, aided by the academia-practitioner inputs and acknowledged the role of SMUS glasses in generating the outputs and understandings of the EKW as a 'dynamic space of practice'. The participating practitioners included representatives from the Central Inland Fisheries Research Institute (CIFRI), the leading scientific advisor to the working mechanism in the region, and three *bheris*—the state-owned Nalban Fisheries, the cooperative Baro Chaynavi Matsyajibi Samabay Samiti and the private Jhagrashisha Fishery, where the field works were conducted.

The students shared their critical understandings of the space derived from the SMUS glasses, i.e. the application of ethnography and visualisation tools that were explained by the academic team in the classroom sessions as essentials of research, explaining how it helped them to perceive the wetlands as a socio-ecological space embedded with deeper complexities that are often overlooked in mainstream conservation debates. These observations were developed through the various visits to these individual fisheries between March and August 2023, where the students learnt theories and hands-on application of the methodology by being exposed to a variety of ethnographic methods (such as participant observations, key informant interviews, focus group discussions), complemented by visualisation techniques (such as photo-video coverage, illustrations and sketches) while exploring everyday socioecological realities shaping the wetlands. They discussed how they had prepared their reflections through the ethno-graphy approach, where their new knowledge of the space from the SMUS glasses had been presented through graphical illustrations (photography, videos, sketches, illustrations, etc.).

The workshop focused on three themes of urban sustainability: a) Biodiversity, b) Pollution and c) Livelihoods dotting this wastewaterscape, perceived by the students before and after the application of the toolkit through various innovative and interactive means of engagement. The presentations were based upon their organic reflections of and understandings about the space derived from the first two stages of the PEIP—(1) Data collection and (2) Analysis. The students expressed their understandings, analysis and hopeful suggestions based on their work in the case sites, discussing how the toolkit helped them in 'seeing' the space in a different lens.

> Our wetland is not one space, but a hub of multi-layered and heterogeneous entities that many locals are not even aware of…

The above quote was one of the many reflections that were collectively echoed by the students during their presentations. It was not one but every student's opinion that they presented in front of the audience. The workshop provided a unique opportunity for the multiple stakeholders of the wetlands, the practitioners, to come together and reflect upon these deliberations, individually and also in joint effort with their peers—each *bheri* representative and other practitioners optimising the opportunity of mutual exchange through this academia-practitioner collaborative workshop, and sharing critical feedback on the toolkit, its applicability and challenges of implementation. The first session consisted of the theme-wise presentations, developed by the students in three groups, with each group comprising two mentors from IIT Kharagpur and Disappearing Dialogues and 5 to 6 students. Group 1, whose theme was biodiversity, presented the role of four agents of the wetlands flora and fauna—fish, water hyacinth, worms and bacteria—that they had identified during the interviews with practitioners during field visits, along with observations in and around their homes, the road to school and discussions with community members that were either simplified or overlooked as part of the EKW biodiversity. They presented a photo-illustration, where they discussed the complex 'cycle of activity' between the four agents in increasing the socio-economic and ecological valuation of land and water in the wetlands and nudging towards further ecosystem-based livelihood strategies in the region by local inhabitants.

Group 2 presented a video narrative and a sketch on pollution in the wetlands, where they categorised the different aspects of pollution in the region and shed light on the challenge of inorganic waste and a lack of awareness on plastic waste management in this space. During their interactions with *bheri* officials, they had learnt about the increasing challenge of waste disposal in the region, which despite being a waste recycling resource was struggling due to the changing composition of waste, i.e. from organic to plastic, in the region. The students then further transected the wetland space collecting evidences of plastic waste management problems across the region, capturing it through interactions with community members and through the mobile devices provided to them. Their video narrative provided an engaging narrative of the wetlands' natural functioning arteries being choked from the deposition of non-bio-degradable waste from the city as well as locals, and the lack of policy recommendations to tackle this in a Ramsar protected site.

Group 3 presented an illustrative representation of livelihoods in the wetland region, which was developed through the application of the SMUS Toolkit during and after the field visits. During the field visits the students had interacted with the different *bheri* officials and members to learn about the types of livelihoods practiced in the space, along with associated information on wages, work hours, productivity, quality of life, etc. They further carried on these conversations in and around their homes, roping in community members engaged in both *bheri* and non-*bheri* services, which they classified as fishing and farming activities with further sub-classifications, and compared it to the practices from their case sites. The students also classified their information from a gender perspective, identifying the various roles of women in the region, their wages, work hours and the reason for appropriation of certain jobs for women—such as cleaning the premises and pond of water hyacinths, washing the harvested fish, etc., as it required less physical effort and gave the women more time to pursue their household jobs. Group 3 presented their findings through another ethno-graphy output—a hand-drawn bird's eye view map of the wetland space, showcasing a non-human perception of the connectivity of lives and livelihoods in the EKW.

> *The wetland does not exist to only serve the urban, but to exist, it needs the urban.* (Prianka Mistri, Bamanghata High School)

The above quote by Priyanka, one of our PEIP students is a reflection of the students' changing perceptions towards the city, which was previously only seen in the image of an extractor. This wouldn't have been possible without their exposure to the toolkit, which helped them explore the city-wetland relationship beyond the normative understanding. In the second session the practitioners commended these innovative methods and the conceptual clarity of the young students in understanding the place-based qualities of the region and its people. They also praised the collaborative academia-practitioner platform crafted by PEIP, which not only offered them the opportunity to interact with the students and academia but also their inter-intra wetland agential colleagues. They lauded the SMUS glasses as a necessary apparatus in examining the challenges of the EKW, while briefly reflecting on the need to cross-examine and validate the perceived solutions in this dynamic space of practice in regards to allied challenges and long-term eco-environmental efficacy. The key outcome of the workshop was the unanimous agreement among practitioners to work with the students in co-exploring the EKW using the SMUS Toolkit.

5.5 PEIP-ing Through the SMUS Lens: (Re)adjustments and Reinvigorations

A flurry of debates continued among the internal team members on the inclusivity of the term 'practitioners' itself. Despite trying to blur the boundaries between academia and practice, we found ourselves already using separatory terminologies in identifying the different wetland groups. A section of the academic team had doubts about whether the training sessions were adequate to help students understand the potential and possibilities of the glasses, while another section believed that the students, who were inhabitants of the space and intertwined with its complex socio-ecological assemblage, were already youth practitioners of the space by themselves. The term 'practitioner' could not only be limited to the *bheri* officials and other stakeholders of practice but also reflected upon the deeper connections existing within and outside the community. The students, many of whose parents work directly in the *bheris* or are attached to its provisioning services, carry on the region's identity and legacy through their perceptions and actions, transmitting them among their next generations. Another pertinent question that came to our mind was on the efficacy of the glasses in this socio-ecological context. Can they be readjusted along place-based specificities? We wanted to scrutinise this aspect further and, to accomplish that, organised our next workshop on methodological implementation in the wetland itself. The setting was informal, as a community-shared temple space was organised where the students would share their findings with experts, talk about the role of the glasses and disseminate their understandings with the local community.

As the students explained in detail to the experts and community members about their field works, assignments and shared their outputs, the academic team kept probing them with methodological challenges, their influence on their new outlook and how they had used them to arrive at certain conclusions. The students shared how they used Internet on their mobile phones to learn more about the wetlands after each training session, as they were keen to know more and get familiarised with the space. They would connect with each other to share notes, pictures and maps, have family discussions and brainstorm about assignments. Many community members came forward and provided interesting feedback to the students that they could focus on in future assignments ranging from the region's nature-based services to climate awareness. The critical assessment at this stage proved necessary, as it helped students in identifying three aspects of the wetland space that they could use to introspect with their toolkit and present to the practitioners in the next meeting—biodiversity, pollution and livelihood practices. After a short stock-taking meeting of the work progress, the students were grouped to work on the respective topics, with mentors from the academic team and DD guiding them at crucial junctures. The teams would explore their notes, revisit the field and critically interrogate and triangulate their data, discussing it with the practitioners.

The practitioners included stakeholders from the government-owned Nalban Fisheries, the cooperative Baro Chaynavi Matsyajibi Samabaya Samiti and the private Jhagrashisha Fisheries. The students presented their works to them in iterative rounds of deliberations and feedback that included oral presentations, photo essays, video narratives, etc., where they clearly explicated the role of the SMUS glasses as an enabler in generating fresh vantage points on the EKW. They discussed how the glasses had helped them in understanding the city's role in the survival of the wetlands, the need for local awareness about pollution, possibilities of alternate livelihoods to supplement income and the importance of multispecies interactions in the wetlands' sustenance. The presentations were appreciated wholeheartedly by the invited practitioners, who applauded the students' efforts in understanding the wetlands beyond its conventional definition and constructing a platform for the different members of the practitioner community to envision a future of fraternity amidst and outside the wetlands. However, they also provided critical feedback and suggestions to some of the solution-oriented assumptions that, they argued, could be corrected or enhanced with more interactive engagements in the field. This prompted the three teams to visit the *bheris* again and conduct further investigations, but now with more focused agendas. Group 1, that worked on the biodiversity of the wetlands, visited the Jhagrashisha Fishery, whose leaseholder Dr. Sasidulal Ghosh, a scientist and former deputy director of the Government of West Bengal Department of Fisheries, and a representative of the statutory body of the East Kolkata Wetlands Management Authority, provided the students with in-depth understanding of varied species of flora and fauna. He appreciated the students' attempts in identifying four major biodiversity contributors in the wetlands, namely, coliform bacteria, fish, water hyacinth and worms, all of which help in the region's sustenance and can be used as part of ecosystem-based livelihood purposes among the community. Group 2, working on pollution in the wetlands, visited the Nalban Fisheries, which is located at the closest periphery of the city and explored pollution-based challenges affecting the space. An important aspect identified in their conversation was the limited awareness on plastic waste among wetland habitants, who are unable to find space to dispose it. Most of the members erroneously deal with it by either burning or burying it, which affects the soil and air quality of the region.

The academic team documented these findings as a crucial area demanding policy intervention through awareness and sensitisation programmes and a practical roadmap. Group 3 focused on livelihood in the wetlands, with the cooperative Baro Chaynavi, where they investigated alternate livelihood strategies for community women through nature-based resources such as fish pickles, medicines, ornaments and furnishings from tree trunks, leaves, flowers, etc., that can supplement their income along with ensuring wetland sustainability. Gobinda Sardar, the secretary of the cooperative, acknowledged that the students' assignment could actually provide classification of feasible action plans to economically and culturally empower the women in the region.

These multiple and dynamic interactions were crucial at this stage, because it allowed all the actors to re-evaluate their strategies in the deployment of the toolkit. The SMUS glasses were essential in understanding and internalising the everyday spatialities dotting the EKW, both for the students and academicians. When the SMUS glasses were taken to the local practitioners, they not only co-explored the wetland, capturing finer materialities and intangible elements but also came up with critical suggestions that helped in readjusting the glasses and providing fresh, tangible insights that would actually curate sustainable pathways for the wetlands, the city and their interconnections.

5.6 Ethno-graphy as a Participatory, Agency-Induced Disseminative Tool

Before the practitioner workshop, in June 2023, the PEIP students received an invitation to present at the 'Youth for Climate' programme, organised by the Living Waters Museum at the American Centre Kolkata, where they showcased their illustrations and the 3D model of the Nalban Fisheries. This event was organised at the early stages of the project, with the ongoing training sessions and the students juggling their project assignments alongside their school curriculum. The invitation to attend a roundtable discussion on protection of the wetlands was extended to selected students from the PEIP project. Despite the short notice and limited time for preparation, the students took onus of their roles as ethno-graphers and provided an insightful discussion on the importance of the EKW, its complex mechanisms and the co-dependent relationship between the city and her wider ecological infrastructures that they had learned through the use of the SMUS Toolkit. They discussed their role in this project, various research methods learnt and applied by them, and shared their ethno-graphy outputs to engage the audience in a lively dialogue on the everyday spatialities of the EKW.

After this programme, the next event where the students shared their findings was the local practitioner workshop, which proved to be an eyeopener for the PEIP team (Sect. 3.3), allowing the PEIP team members to reconceptualise their motivations and objectives according to the practitioners who had worn the glasses. Data had been collected, analysed, implemented and then reconceptualised to co-curate a more holistic but nuanced understanding of the EKW. The need to find an innovative pathway to disseminate these understandings to an audience was prudent, considering the momentum reached after the iterative field visits and academic-student-practitioner co-exploration of the wetlands. Each group (including the local practitioners from the three *bheris*) worked on the previous themes, but with a more nuanced edge, tying any loose threads together and developing an innovative-engaging strategy for dissemination.

The stage was set for the final workshop or exhibition at the Modern High School International, Kolkata, jointly organised by the PEIP team with the *Living Waters Museum* as part of their *Jol Jyanto Kolkata* event, where students, teachers and officials from several esteemed institutions would partake to understand the complex wastewaterscapes of Kolkata. The PEIP team had developed performances based on the three topics: a) 'From city, to nature and beyond'—a *patachitra* play narrative on wetland biodiversity, b) 'A trip to the wetlands'—a play and visual narrative on the challenges of pollution in the wetlands, and c) 'From marshes to markets'—a portrayal of the methodological techniques used to explore the livelihood dynamics in the wetlands.

The *patachitra* play demonstrated the story of a crow from the city who chanced upon the wetlands and the various species living there such as a tree, fox, fish and coliform bacteria, with whom she engaged in an insightful discussion on the city-nature relationship, its history and mutual existence. The students enacted the different characters and developed a *patachitra*[3] displaying the story of these multispecies interactions and their perception of the city-wetland relationship over the years (Fig. 5.7). It depicts the coliform bacteria, derived from sewage as taking the centre-stage in the picture as well as the conversation with fish, animals, trees and the urban structures in the backdrop, showing the existing and manufactured connections between the city and the EKW's socio-ecological presence. The next presentation, a methodological play aimed at recreating one of the scenes from the fieldtrips, exposed the audience to the methods used by the students to collect data, showcasing how the students engaged in discussions with the *bheri* officials and experts, about the livelihood challenges faced by the locals and possibilities for more livelihood opportunities by using the city-nature interface in the future. They also developed a 3D model of the livelihood scenarios and opportunities present in the region (Fig. 5.8), along with a Z card, narrating their own PEIP journey of (un)learning the dynamics of their inhabited space. In the last presentation, a play on wetland pollution, the students played the roles of different types of pollution such as air, water and soil, as well as sound pollution, and depicted its catastrophic effects on the wetlands while discussing solutions to tackle it in future. The students presented a compassionate plea to the audience as to why and how this urban ecological heritage could be protected and safeguarded, and how ethno-graphy not only enabled them to know the wetlands better but also acted as a participatory, agency-induced disseminative tool crafting long-lasting cognitive impact.

These ethno-graphic presentations were not only developed by the students but by the practitioners as well, who had provided the students critical feedback on their previous presentations and further enhanced their knowledge, vis a vis, through continued interactions. All the presentations provided an interactive and engaging understanding into the wetland's everyday reality, often overwhelmed by the constant discourses on oppressive statecraft practices and a homogenous understanding of its conservation. The PEIP project had opened up avenues for awareness, collaboration and stakeholder fraternisation, where the different actors involved in the management of the wetlands could envision the common aim of protecting this socio-ecological resource, while appreciating the varieties of complex socio-cultural dynamics embedded in the historicity of the region. The students had succeeded in applying the toolkit and disseminating their coproduced knowledge through 'ethno-graphy', where the audience, both city and wetland dwell-

---

[3] A *patachitra* is a traditional, cloth-based scroll painting with folktales inscribed in it, popular in eastern India. Natural elements are used for this distinctive style of painting on the *pat* (long cloth), and it is presented via folk songs (Chatterjee et al., 2021).

Fig. 5.7 The *patachitra*. (Source: PEIP student team, under the guidance of Swarnadeep Bhattacharya, IIT Kharagpur)

ers, felt interested as well as comfortable in learning more and new things about the wetlands.

5.7 Conclusion

The inception of this PEIP began at a time when the wetland-city dichotomy prevailed and overshadowed rational approaches to the protection and co-management of the EKW as a contested space. Reeling under the challenges of land grabbing and conversion into real estate and sporadic bouts of activism, the ongoing perception of Kolkata served as an antithesis to that of the EKW. Could hope be curated for this otherwise contested ecoscape through a systemic assertion of engagement and interaction between diverse yet connected worlds of academia and practitioners? What would be the pedagogy for this? Who would lead the venture?

While the idea of the EKW as a natural and cultural heritage and the need to protect it has been eminent among various sectors of the city and wetlands, such as the *bheri* practitioners, state officials and environmental activists, the actualisation of this process always had been a discourse of different motivations. There had been a link missing between the voices of the inhabitants, from both the city and the contested space, with the former being mostly unaware of

Fig. 5.8 Another 3D model depicting the co-transected area. (Source: PEIP student team, under the guidance of Swarnadeep Bhattacharya, IIT Kharagpur, and Sasidulal Ghosh, leaseholder Jhagrashisha *bheri*)

the regions vitality and the latter being victimised and uninformed of the city's contribution in the wetland's sustenance. The deployment of our PEIP design not only facilitated the (re)interpretation of the EKW's complex spatio-cultural dynamics from the classrooms to the field by agentialising the students as the inhabitant critical thinkers of the region but also as the guardians of this natural cultural heritage of the region.

The spatial methods allowed the wetland students and other inhabitants to fill up this void and use their newly gained perspectives to generate place-based understandings and the need for crafting a collaborative platform of future engagements. The SMUS Toolkit and the ethnographic exercise stimulated their understandings and reflections upon realities, optimising on opportunities of 'co-creation of knowledge through interaction (in the project team, illustration), and a curriculum grounded in students' (and practitioners') interests and experiences' (Bartlett, 2008).

The role of transdisciplinarity emerged as the most tangible way forward, in providing a platform to the unheard or suppressed voices through the innovative and interactive tool of ethnography. The wetland students, who began this project with an apprehensive view of the city, learnt about the collaborative role of city thinkers and native indigenous practices in the historical evolution of the space, signifying the need for similar exchanges in future for the region's continued sustenance.

One of the most important outcomes of this PEIP was witnessing the evolving dynamics between the students and the practitioners of the space, coming forward to forge an alliance of involvement-engagement and accountability towards their inhabited space, while reflecting upon the declentionist narrative of urban oppression embedded in their minds and forging out ways to overcome it. The role of the academic team was crucial in terms of acting as both presenters of the Toolkit and facilitators of the immersive-innovative-interactive exercise. In turn, the academic collaboration with the practitioners could also be forged in more meaningful and impactful ways through the students' roles in explaining the imperative of the stewardship framework.

Hence, transdisciplinarity ended up transforming the EKW from a contested into a collaborative space—a necessary and crucial step in safeguarding the natural and cultural heritage of Kolkata.

Acknowledgements We would like to acknowledge Dr. Anuradha Choudry, Assistant Professor, Department of Humanities and Social Sciences, IIT Kharagpur for co-leading this project and providing her critical inputs during the project design and execution phases. Furthermore, we thank *Disappearing Dialogues*—the creative collective that was involved in organising the school events. Likewise, we are grateful to Sukrit Sen from Living Waters Museum as well as to Raktima Ghosh and Swarnadeep Bhattacharjee, the IIT Kharagpur students who trained the young adults to get acquainted with and use the 'ethno-graphic' methodology. Last but not least, we would like to express our heartfelt gratitude to the wetland community who supported us in successfully running the project and co-participated in this venture.

References

Bandopadhyay, A., Ghosh, R., Mukherjee, J., & Pathak, S. (2022). From the shabars of the Indian Sundarbans: Everyday empirics through photography. *Coastal Studies and Society, 1*(2–4), 123–139.

Bartlett, L. (2008). Paulo Freire and peace education. In *Encyclopaedia of peace education*. Teachers College, Columbia University.

Bhattacharyya, D. (2018). *Empire and ecology in the Bengal Delta: The making of Calcutta*. Cambridge University Press.

Bhattacharya, S., Mukherjee, J., Banerji, D. and Chatterjee, S. (2024). *(En)countering an urban riverscape: Ethno-graphic explorations on the Adi Ganga*. Cities. Elsevier.

Chatterjee, A., Bhattacharya, S., & Mukherjee, J. (2021). Nature, culture and oriental heritage: Ethnographic explorations on Patua and Chhau communities of Bengal. *The Oriental Anthropologist, 21*(2), 253–271.

Dembowski, H. (2000). *Taking the state to court: Public interest litigation and the public sphere in metropolitan India*. Oxford University Press.

Freire, P. (1990). *Pedagogy of the oppressed*. Continuum Press.

Ghosh, D. (2005). *Ecology and traditional wetland practice: Lessons from wastewater utilisation in the East Calcutta*. Wetlands Worldview.

Kundu, N., Pal, M., & Saha, S. (2008). East Kolkata Wetlands: A resource recovery system through productive activities. In *Proceedings of Taal 2007: The 12th World Lake Conference* (pp. 868–881). http://www.moef.nic.in/sites/default/files/nlcp/WLC-Report.pdf. Accessed 29 Mar 2019.

Mukherjee, J. (2015a). Beyond the urban: Rethinking urban ecology using Kolkata as a case study. *International Journal of Urban Sustainable Development, 7*(2), 131–146.

Mukherjee, J. (2015b). Sustainable flows between Kolkata and its peri-urban interface. In A. Allen, M. Swilling, & A. Lampis (Eds.), *Untamed urbanisms*. Routledge.

Mukherjee, J. (2020). *Blue infrastructures: Natural history, political ecology and urban development of Kolkata*. Springer Nature.

Mukherjee, J., & Bhattacharya, S. (2023a). *Understanding nature and societies through ethno-graphy: A transdisciplinary perspective*. Springer Nature. Video Publication. https://doi.org/10.1007/978-981-99-6725-4

Mukherjee, J., & Bhattacharya, S. (2023b). Ecology, well-being, and community resilience: Lessons from deltaic South Asia. In S. Chetri, T. Dutta, M. K. Mandal, & P. Patnaik (Eds.), *Understanding happiness: An explorative view*. Springer.

Mukherjee, J., & Véron, R. (2023). Urban environmental governance: Historical and political ecological perspectives from South Asia. In E. O'Gorman, W. S. Martín, M. Carey, & S. Swart (Eds.), *The Routledge handbook of environmental history*. Routledge.

Mukherjee, J., Bhattacharya, S., & Bose, L. (2022). Heritage or basic human rights? Politics of environmentalism surrounding the Adi ganga in Kolkata. In S. Pattanaik & A. Sen (Eds.), *Contested landscapes and regional political ecologies of environmental conflicts in India*. Routledge.

Myint, T. (2018). *The long read: From the Anthropocene to the Urbanocene: Understanding Asia's rural out-migration and global climate change*. https://theasiadialogue.com/2018/12/14/from-the-anthropocene-to-the-urbanocene-understanding-asias-rural-out-migration-and-global-climate-change/. Accessed 23 Oct 2023.

Palme, M., & Salvati, A. (2021). Introduction: Anthropocene or Urbanocene? In M. Palme & A. Salvati (Eds.), *Urban microclimate modelling for comfort and energy studies*. Springer.

Ramsar Convention on Wetlands. (2018). *Scaling up wetland conservation, wise use and restoration to achieve the Sustainable Development Goals*. https://www.ramsar.org/sites/default/files/documents/library/wetlands_sdgs_e_0.pdf. Accessed 23 Oct 2023.

Schott, D. (2004). Urban environmental history: Water lessons are there to be learnt? *Boreal Environment Research, 9*, 519–528.

West, G. (2017). *Scale: The universal laws of life and death in organisms, cities and companies*. Hachette.

Jenia Mukherjee is an Associate Professor of History of Ecology and Environment at the Department of Humanities and Social Sciences of the Indian Institute of Technology Kharagpur. Her interest lies in urban studies and transdisciplinary waters. In the book *Blue Infrastructures*, she proposed and applied historical urban political ecology (HUPE) to explore the more-than-contested wastewaterscapes of Kolkata, discussing its viability in forging academia-practitioners dialogues and actions. Recipient of the World Social Science Fellowship (2013), the Carson Writing Fellowship (2018) and the Nippon Foundation Fellowship (2020) for advancing urban ecological research, she is an active SMUS partner, investigating multiple projects from India. Together with Shreyashi Bhattacharya, Mukherjee has experimented with the efficacy of the 'ethno-graphic' approach in capturing (urban) environments of South Asia. Their latest output on 'ethnography' is available as a Springer video from 2023.

Shreyashi Bhattacharya is a Senior Research Fellow of the Rekhi Centre of Excellence for the Science of Happiness at the Indian Institute of Technology Kharagpur, where she concluded her PhD on socio-ecologies of the Adi Ganga riverine system by addressing the history and resilience of the Adi Ganga, a heritage river flowing through the dense urbanscapes of Kolkata. In this study she used the transdisciplinary lens of 'ethnography' to study the river's 'hydrosociality'—historically complex and dynamic socio-ecological assemblages. Shreyashi has received the prestigious Writing Urban India Fellowship, Huntington Fellowship and Living Waters Museum Fellowship for her doctoral thesis.

Sasidulal Ghosh is the leaseholder of Jhagrashisha Bheri in the East Kolkata Wetlands. He was the former Deputy Director of the Department of Fisheries at the Government of India. He is passionate about inland fishing experimentation and sewage-fed fisheries. He is also a representative of the East Kolkata Wetlands Management Authority and has been a key member in the Wastewater User Association and Fish Producers' Association.

Open Access This chapter is licensed under the terms of the Creative Commons Attribution 4.0 International License (http://creativecommons.org/licenses/by/4.0/), which permits use, sharing, adaptation, distribution and reproduction in any medium or format, as long as you give appropriate credit to the original author(s) and the source, provide a link to the Creative Commons license and indicate if changes were made.

The images or other third party material in this chapter are included in the chapter's Creative Commons license, unless indicated otherwise in a credit line to the material. If material is not included in the chapter's Creative Commons license and your intended use is not permitted by statutory regulation or exceeds the permitted use, you will need to obtain permission directly from the copyright holder.

6

Hybrid Use of Spatial Methods in Transdisciplinary Urban Sustainability Studies: Perspectives from Bangkok

Jakkrit Sangkhamanee, Piyathep Tanmahasmut, and Poramin Watnakornbancha

Urban sustainability studies require multidimensional and transdisciplinary approaches that integrate social, economic, and environmental considerations. Spatial methods, among many others, can assist in examining these intricate interactions and comprehending the spatial patterns, dynamism, and relationships underlying urban sustainability. In this chapter, by engaging with literature from different case studies and disciplines and reflecting on our experience in Bangkok, we will examine the applicability and appropriateness of recent research on spatial methods for urban studies in relation to diverse stakeholders, contexts, and issues of urban sustainability. Spatial thinking is especially significant when we consider the methodology for studying the dynamism and multiplicity of the city within the context of a society in transition (Logan, 2012). The application of transdisciplinary approaches and methods to spatial studies can be one solution for addressing such complexities.

We argue that spatial methods for urban studies have recently shifted primarily from a focus on territorial space to a greater emphasis on relationships within and between spaces. This evolution is exemplified by the methodologies used to comprehend the complexities and dynamics of urban challenges and sustainability. In this contribution, we attempt to depict such transformative use of spatial methods by highlighting some urban research from Bangkok that demonstrates the hybrid use of methodologies, as well as an experimental activity to investigate new methodological possibilities for capturing the nuances of urban lives (human and non-human) and materiality.

To illustrate the shift and hybridity in spatial methodologies, we first explore recent spatial methods in urban studies. These methodologies cover an array of techniques and approaches involving the analysis, interpretation, and visualisation of spatial data to comprehend the dynamics, complexities, and sustainable conditions of cities. These techniques and approaches are helpful for making use of spatial data such as maps, satellite imagery, and geographic information systems (GIS) to investigate patterns, relationships, and processes within urban territories. In addition, influenced by the mobility and interpretative turns in social science (Sheller & Urry, 2006; Sivado, 2020), spatial methods also include

J. Sangkhamanee (✉)
Department of Sociology and Anthropology + SMUS, Chulalongkorn University, Bangkok, Thailand
e-mail: jakkrit.sa@chula.ac.th

P. Tanmahasmut
Chulalongkorn University, Bangkok, Thailand

P. Watnakornbancha
Ari Ecowalk, Bangkok, Thailand

non-visual data such as the ones brought about by an ethnography of daily spatial relationships in the city. Last but not least, recent ontological and affective turns (Clough & Hallry, 2007; Gullion, 2018; Holbraad & Pedersen, 2017; Sivado, 2015), socio-atmospheric conditions, non-human and multispecies relations, as well as assemblages of material multiplicities and itineraries beyond a human-centred approach are crucial factors in understanding urban conditions and sustainability (cf. also Amin, 2014; Anderson & Holden, 2008; Brenner et al., 2011; Farias, 2011; McFarlane, 2011a, b; Thibaud, 2015; Thrift, 2014). Indeed, contemporary spatial methods for urban studies have evolved into more hybrid relationships between space and humans and non-humans, as well as material and immaterial co-creations. We will return to this approach with a case study from Bangkok at the end of the chapter.

But to begin with, let us discuss some of the underpinning reasons why urban studies employ spatial methods for their studies. First, spatial methods can be utilised to comprehend spatial patterns. As we will see from case studies below, multi-disciplinary researchers can identify and analyse spatial patterns in urban areas, such as the distribution of land use, population, and infrastructure, using spatial methods. This aids comprehension of the spatial organisation and operation of cities (see Pinfold & Mokhele, 2019; Sands, 2009; Turner, 2011). Additionally, spatial methods can be utilised to investigate the interrelationships and interactions between various urban system components. For instance, they can be used to analyse the effect of transport networks on accessibility, the effect of green spaces on well-being, and the effect of social segregation on urban dynamics (see Checker, 2011; Hawken et al., 2021; Pedro et al., 2019; Wei et al., 2018). In addition, spatial methods can facilitate the visualisation and communication of urban research findings. Spatial methods have facilitated the development of visual tools that can represent and convey urban phenomena. Maps, spatial models, and data visualisations assist researchers and policymakers in comprehending and effectively communicating spatial information (see Di Chiro, 2018; Huang & London, 2016; Timur & Getz, 2008). And lastly, spatial methods can support the evaluation of sustainability and the planning of interventions. By analysing spatial indicators and relations and measuring and understanding environmental, social, and economic conditions, spatial methods can contribute to the evaluation of urban sustainability. They support evidence-based decision-making in urban planning and policy by identifying intervention areas and potential sustainable development strategies (see Cadieux, 2008; Pinfold & Mokhele, 2019).

We highlight three spatial methods in our brief review of Bangkok case studies below: Geographic Information Systems (GIS), spatial analysis, and urban ethnographies. These three methodologies are applied to a variety of urban research topics, including city flood, transportation planning, land use analysis, environmental assessment, and urban resilience, either independently or in combination. We argue that incorporating spatial methods into urban studies allows researchers to gain a better understanding of the spatial dimensions of urban processes and dynamics. Spatial methods, as illustrated in the case below, provide insights into the interactions between the city's physical and social aspects, assisting in the resolution of sustainability issues and informing policy interventions that promote more efficient, equitable, and sustainable urban environments.

Following our review of existing literature and spatial research on Bangkok's urban resilience, we then propose our own case studies in the latter section, which are based on our project to connect city people with their environment. We propose that our case study sheds light on additional, experimental spatial methodology, namely, human sensorial and affective methods, for comprehending the urban environment and its relationships to its inhabitants. While the goal of this chapter is not to provide a comprehensive review of existing spatial methods, we hope that the perspectives from Bangkok can shed some light on the relevant urban issues and how some selected methodologies can assist in dealing with the problem or finding solutions to public concerns.

6.1 Bangkok Urban Resilience and Spatial Methods

Bangkok, like many regions of the world particularly in Asia and Africa, is experiencing the challenging effects of rapid urbanisation, which include overpopulation, informal settlements, inadequate infrastructure, and strained public services. In addition to rapid urbanisation, the city is facing environmental issues such as pollution, a lack of urban wilderness, biodiversity loss, and insufficient waste management (Yamashita, 2017). The situation worsens when rapid urbanisation and environmental degradation are combined with climate change and resilience concerns. The effects of climate change, including rising sea levels, extreme weather, and heat islands, pose a threat to the city (Sangkhamanee, 2021; Thanvisitthpon et al., 2018). In other words, Bangkok urban sustainability faces numerous challenges in a variety of urban phenomena and city resilience. Some common challenges observed in the city include, but are not limited to, rapid urbanisation, climate change, flood and resilience, green space and land use change, social inequality and housing, environmental degradation, and governance and participation (see Ratanawaraha, 2016; Sintusingha, 2011).

If we place Bangkok's sustainability against United Nations' Sustainable Development Goals (SDGs), the picture becomes even clearer. SDG11 focuses on making cities and human settlements inclusive, safe, resilient, and sustainable. The goal emphasises the importance of creating urban environments that promote quality of life for all residents, enhance social cohesion, and contribute to environmental sustainability. It addresses various aspects of urban development, including housing, transportation, urban planning, cultural heritage, disaster resilience, and access to basic services. Bangkok still faces many urban challenges to achieve the above goals.

Various spatial approaches are specifically suitable for investigating different problems and applying them in different situations to achieve the SDGs. However, due to the complexity and dynamism of urban conditions, in many cases, spatial methods must be refined and combined with other research methods, as well as involving transdisciplinary partners and stakeholders, to reflect the conditions of urban issues and sustainability and collaborate on transdisciplinary research practices. In this section, we discuss the application and appropriateness of spatial methods utilised by transdisciplinary partners to address a variety of issues in various contexts. Based on our literature survey on Bangkok's urban challenges and resilience-based policies, the purpose of this section is to shed light on the critical evaluation and reflection of how spatial methods can be used to understand and improve urban sustainability as well as to shape research practises in transdisciplinary collaboration. The methods under discussion, however, are not exhaustive but are based on the available practises and publications observed in academic and policy-focused research in Bangkok. We highlight three spatial methodologies, namely, geographic information systems, spatial analysis, and ethnography, to reflect on their capacities and appropriateness in applying them to Bangkok's urban research.

6.1.1 Geographic Information Systems (GIS)

Utilising GIS to map and analyse urban sustainability is a crucial aspect of urban spatiality-related research (Giordano, 2023). Typically, the GIS method is appropriate for visualising the larger study region. However, in many instances, the method is used in conjunction with other techniques to obtain a closer look at details or to illustrate the area's dynamic nature.

From our survey, several studies have employed spatial methodology to investigate urban resilience in the context of Bangkok. For example, one application of GIS in anticipating future urban development is elucidated in Thaiutsa et al.'s study (Thaiutsa et al., 2008), which aims to gather information for identifying suitable sites for future green spaces. The research adopts an urban green space analysis approach, utilising data from the Bangkok Metropolitan Administration (BMA) and the Silviculture Department at Kasetsart University. The research partners here thus consist of both government administrative authority and academic institutions. Its research variables encompassed park, tree, population, plant, flower, and area in Bangkok. The study presents the predominant tree species in Bangkok, emphasising the need for remaining tree information for effective urban green infrastructure development. Challenges related to tree preservation during dry seasons are acknowledged. GIS analysis illustrates the sustainable urban planning initiatives of the BMA, manifesting in the addition of new parks.

Another example is the case where the water engineering team at the Asian Institute of Technology (AIT) utilised GIS to devise a cartographic representation of Bangkok's groundwater resources (Neupane et al., 2022). This study aimed to evaluate the resilience of groundwater in Bangkok and its environs amid the dual challenges of climate change and urban development. Following the collection and analysis of climate data, the research team generated a climate model and a land use change scenario to project future trends in groundwater recharge and levels. Based on GIS dataset, the outcome of this research served as an indicator to construct a groundwater resilience map for the area, revealing a severe threat to Bangkok's groundwater resources from the impacts of climate change and land use changes. This example illustrates how GIS aids researchers in forecasting prospective scenarios of urban resilience through the integration of historical land data and statistical information.

In exploring the relationship between green areas and land use change, Kamal et al. (2017) employed the Multi-Layer Perceptron Markov Model to detect and analyse land use changes from 1994 to 2030. The study identifies a decrease in vegetation and cultivated lands, attributing it to rapid urbanisation. The study emphasises the lack of a sustainable township plan to address urban disasters and the diminishing green areas from extensive parts to urban cores. In addition, Malcolm (2002) addresses the environmental stress resulting from rapid urban expansion in Bangkok, utilising GIS and remote sensing to conduct potential environmental stress analysis. The research aims for sustainable urban environment management and integrates imagery data within the GIS environment, combining it with ground-based information. The combined methodologies enable temporal analysis, revealing percentages and rankings of environmental stress potential areas.

In conjunction with GIS, spatially comparative analysis serves as an indispensable tool for understanding urban resilience. For example, Monprapussorn and Phoung Ha (2021) conducted comparative case studies exploring the integration of land use, climate, and water resources for enhancing resilience in both Bangkok and Hanoi. From their study, they project climate change effects by 2050, considering patterns of rainfall, predicted land use changes, and historical data. Both cities are anticipated to experience increased temperatures due to rapid urbanisation and reduced agricultural areas. In another comparative analysis, Posuk et al. (2018) contribute to persistent urban problems with a GIS-based, comparative analysis of temporal planning and land use changes in Bangkok over a 21-year period. Geological images and city planning data were gathered from US geological organisations and Thai city planning authorities, with GIS data downloaded from the BMA GIS Center. Results indicate a dramatic increase in urban areas, prompting the proposal to control residential zones and halt urban expansion in rural areas to mitigate flooding. It is found that GIS can be helpful in this situation by offering data for comparative analysis that goes beyond a simple spatial comparison between two or more locations but also allows for a comparison between various temporal scales and dimensions, which can be used as a foundation for future policy planning and decision-making.

From our survey, we also found that GIS also plays a pivotal role in urban planning solutions, as demonstrated by Iamtrakul et al. (2017), who used geospatial analysis to propose a hybrid canal-rail network for transit-oriented development in Bangkok. Their study draws upon commuter data provided by governmental organisations, advocating for the integration of water transportation with skytrain or metro systems to address mass transit issues. Another example is Ahamed-Broadhurst's (2018) research, which utilises GIS for mapping and data analysis to explore the relationship between Buddhist temples and canals along the Chao Phraya River in Bangkok. Aerial photos from 1932 are compared with satellite images from 2016, revealing a significant reduction in the number and length of canals. The study highlights the reliance of Thai people on rivers and temples, illustrating the urban development patterns in Bangkok.

In summary, GIS is a valuable spatial technique that can be employed independently or in conjunction with other techniques and datasets to depict the current state and dynamic nature of urban conditions and challenges. This facilitates collaboration among various stakeholders and partners to generate solutions and insights for urban resilience. Furthermore, it can serve as a helpful tool for conducting comparative studies to offer a comparative viewpoint on relevant issues that inform the later stages of policy design and decision-making process. However, there are limitations to the use of GIS in studying urban resilience. One significant limitation is the quality and availability of data. In many cases, there may be a lack of reliable and up-to-date spatial data. This data gap can lead to inaccurate analyses and misguided resilience strategies. Another challenge is the complexity and cost associated with GIS technology. Effective use of GIS requires specialised skills and resources, which may not be readily available in all urban planning departments or among stakeholders. Furthermore, GIS analyses often fail to capture the qualitative aspects of urban resilience, such as community networks and social cohesion, which are crucial but difficult to quantify and map. Lastly, there is a risk of oversimplification in GIS models, which may not fully account for the dynamic and interconnected nature of urban systems, potentially leading to incomplete or misleading conclusions about urban resilience.

6.1.2 Spatial Analysis

For GIS to function optimally, it must be utilised in conjunction with another spatial method. One potential method that can supplement GIS in comprehending urban resilience is spatial analysis. Spatial analysis is a set of techniques appropriate to examine and understand spatial patterns and relationships. This can include identifying clusters of social phenomena, examining spatial autocorrelation, and exploring spatial dependencies. Typically, spatial analysis is not only utilised for urban research projects, but it can be enhanced through collaboration with another method and analysis. There are some studies using spatial methodology to explore urban resilience that were undertaken in Bangkok.

In policy research, Rongwiriyaphanich (2012) investigated the impacts of land policy on the shaping of hybrid rural–urban development patterns using spatial characteristic analysis. The research aimed to understand the potential impacts of land policy on urban resilience. Spatial development patterns in the research were based on a set of characteristics that were connected to degrees of resilience. In this study, two former agricultural areas in suburban Bangkok were chosen based on their diverse spatial development patterns; one is Khlongluang, located in the northeast of Bangkok. Another is Western-Nonthaburi, an area situated in the northern part of Bangkok. These two cases reflect different impacts caused by policies applied in each area. While Khlongluang has undergone a transformation into an industrial area, the Western-Nonthaburi land has been affected by the implementation of a landholding size policy, which shifted towards land use for orchards and residential purposes. For resilience enhancement degree, it can also be differentiated by characteristics of hybrid rural–urban development. This case is an instance of the application of spatial analysis contributing to researchers understanding of geographical areas holistically and comparatively. Such a method can be integrated with other in-depth qualitative methods to empathise with dwellers of these areas.

However, when we shifted our focus from broad policy analysis to more focused urban issues, we discovered that studies investigating Bangkok's resilience frequently touch on problems such as flooding, since this is one of the primary issues that experts and authorities are still trying to prevent and lessen. One example of issue-based, policy research that investigates urban resilience and city flood is Laeni et al. (2019). This study employed frame analysis to understand flood resilience policy development in

Bangkok. Flooding is one of the main problems with urban disasters in Bangkok, as the city is situated near the Chao Phraya River delta. In aiming at spatial analysis, the research team started by looking at how policymakers and experts interpreted resilience policies, as well as their ongoing strategic plans. The frame analysis in this research coded the networks and involved stakeholders in their development. This research differentiates itself from mainstream Bangkok's flood risk management research by focusing on Bangkok's institutional arrangements for flood resilience. Based on the results from people involved with the policy and outsiders, the research can frame the Bangkok Resilience Strategy and provide recommendations for policymakers.

To prevent or reduce urban flood disasters, the spatial method should understand past and current phenomena and adapt to uncertain futures. Another case study about the Bangkok flood is from Sitko and Almuhktar (2017), who conducted research using a morphological approach to understand disaster resilience in informal settlements. The case study is of the riverside informal settlement along the Bang Bua Canal. Apart from morphological analysis, the study also included document reviews and key informant interviews with various stakeholders. The study argued that the traditional urban resilience approach proposes that, to be effective, resilience ought to be built before and after disaster. It was an area for low-income people, which was blamed to be the cause of flooding. In this study, four layers of urban form were used to identify topography, public open space, plots, and buildings. With the morphological layers, the researchers found both socio-economic and governance problems in the neighbourhood. Based on their study, morphographical patterns support urban stakeholders in preparing for the risks of disasters. Moreover, the authority would identify the location of the disaster and the type of risk people or the community face.

Climate change encompasses not only disaster-related events like flooding but also includes changes in natural resources. Various spatial analysis methods have been used to indicate and speculate on climate resilience in the water sector. For instance, Koh et al. (2022) utilised Collaborative Risk-Informed Decision Analysis (CRIDA), which helped researchers analyse and illustrate the problem. The research team collected river flow, salinity, and turbidity data as key metrics to map the framework; they also conducted interviews with Metropolitan Waterworks Authority staff as stakeholders in the study. The next step involved site-specific analysis based on key data from stakeholders. The team analysed different variables with critical thresholds to understand historical system failure occurrences and their relationship. This information helped researchers understand the relationships between the value of different times of the year and the failure-proneness of systems. A stress test was also conducted by adjusting variables for possible future climatic conditions.

The comparison of the seasonal validity of river flow, salinity, and turbidity shows that the Bangkok water supply system is fragile to the effects of climate variability. The benefits of the CRIDA in this project also include helping with decision-making on interventions for climate change adaptation. Furthermore, the study of climate change scenarios also helps to address problems in the system that need to be fixed. However, the research team noted that this methodology might be better suited for cities that are smaller than Bangkok due to the reduced uncertainties associated with a smaller spatial scale (Koh et al., 2022). All in all, urban resilience can be studied by the infrastructure that sustains the city; CRIDA helps develop the water supplementary system and plans to adapt to uncertain climatic change. However, this method is a good example of the collaboration between qualitative interviewing and data analysis, which can transform human information to investigate variables.

As we have seen from above cases, spatial analysis, when applied to studying urban resilience, offers significant advantages, such as the ability to visually interpret complex data, identify patterns and vulnerabilities in urban environments, and integrate diverse datasets (demographic, environmental, infrastructural) for a comprehensive understanding of urban dynamics. However, this approach also faces limitations, including the dependency on the availability and quality of spatial data, which can be scarce or outdated, particularly in less developed areas.

The requirement for specialised skills and resources to effectively utilise spatial analysis tools can be a barrier, and there's a risk of oversimplification in models that may not capture the nuanced, interconnected nature of urban systems. Additionally, spatial analysis tends to underrepresent qualitative aspects like social networks and community resilience, which are harder to quantify and map but are critical components of urban resilience. With these limitations, we now turn to ethnography as one of the qualitative methods that can help filing the limitations of both GIS and spatial analysis.

6.1.3 Ethnography

The last spatial method we highlight here is ethnography. Ethnography, as a qualitative research method, can also contribute to the study of urban sustainability by providing insights into the social, cultural, and political dimensions of urban sustainability. Ethnography can contribute to studying urban sustainability as a spatial method by providing a deep understanding of the social, cultural, and political dynamics of urban environments. By observing and engaging with the everyday practices of urban residents, ethnography can reveal how urban sustainability is negotiated and experienced in different contexts.

Several ethnographic studies have been undertaken in Bangkok, exemplified by Marc Askew's (1999) examination of the management of neighbourhood-level environments in suburban areas. The 1990s witnessed suburbanisation in Bangkok, marked by the development of housing estates by private companies and state organisations. The proliferation of gated communities on the outskirts significantly impacted social structures, leading to the segregation of lower socio-economic classes. Askew focused specifically on the Prachaniwet Project, an affordable settlement initiative by the National Housing Authority (NHA) aimed at addressing housing shortages among low- and middle-income earners. Participating in the weekly coffee club meetings in Prachaniwet allowed Askew to gain valuable insights into the daily lives, circumstances, and challenges faced by residents. The shared social and economic status of Prachaniwet residents facilitated the formation of bonds, contributing to a sense of place and neighbourhood that sustains residents' well-being and addresses environmental problems. Askew's research highlighted the gap between urban authorities and communities while emphasising how Prachaniwet represented the lower middle class's well-being through a developed sense of place and neighbourhood (Askew, 1999).

Erik Cohen's work (Cohen, 2012) offers another example of ethnographic inquiry in suburban Bangkok, employing auto-ethnography to reflect on his experience during the severe floods in 2011. Central Thailand faced unprecedented floods associated with monsoon rainfall and storms, causing widespread damage. Cohen's personal account describes the preparations made, including sandbagging his home and community members building protective walls. The use of auto-ethnography allows Cohen to reflect on his experiences during the flood, contributing a personal perspective to disaster studies.

Ethnographic studies, however, face challenges, often neglecting people within the city while focusing more on remote villages. Sopranzetti (2018) addresses this blind spot in urban anthropology by conducting fieldwork in Bangkok, focusing on motorcycle taxi drivers as key informants. Through participant observation and following drivers to their rural hometowns, Sopranzetti reveals the invisible role of motorcycle taxi drivers in the city's dynamics. These drivers, largely ignored by capitalism, navigate the city's blocked traffic during peak hours, playing a crucial role in urban circulations. Sopranzetti's work demonstrates how ethnography can conceptualise not only social aspects but also political and economic dimensions, amplifying the voices of marginalised groups (Sopranzetti, 2018).

Ethnography extends beyond the study of human relations in social contexts; it can also encompass the study of non-human entities, especially in urban settings. Multispecies ethnography is a kind of ethnography that seeks to understand the complex social and ecological relationships between humans and non-human species in urban environments (Hamilton & Taylor, 2017). This approach can contribute to the study of urban sustainability by illuminating the ways in which human and non-human actors interact and influence each other in ecosystems.

For example, in non-Western urban settings such as Bangkok, a distinctive spatial feature known as *soi* (or alleyway) serves as a connection between residential areas and main roads. Based on more-than-human perspective towards urbanity, Nikki Savvides (2013) conducted ethnographic research on semi-feral dogs inhabiting the alleyways of Bangkok. To immerse herself in the field, Savvides volunteered for a Western-run organisation providing care for stray dogs and cats in Bangkok. This enabled her to observe interactions between people in the alleyways and the dogs. The study sheds light on the challenges associated with the cohabitation of *soi* dogs and people in these communities.

Initially, Savvides labelled the stray dogs in Bangkok's market community as *soi* dogs. One significant concern is the fear associated with these dogs, as some may attack humans and others may carry rabies. Many residents of Bangkok tend to avoid and fear these dogs when encountering them in the alleyways. In contrast, market stall owners, generally belonging to the working class, routinely provide scraps of food to the dogs before closing their shops. Some even develop specific bonds with specific dogs they care for. The practice of feeding the dogs is culturally rooted in Thai-Buddhism beliefs of accruing merit. Moreover, the *soi* dogs play a protective role in communities, guarding against burglars and outsider dogs that intrude into their territories. However, the provision of food to canines contributes to the problem of overpopulation, leading to diseases and contamination associated with their presence (Savvides, 2013). The study of *soi* dogs exemplifies the application of multispecies ethnography to examine hybrid urban ecology in Bangkok. This method allows ethnographers to gain insights into the coexistence of urban animals and humans, revealing the multispecies relationships shaped by socio-economic and cultural factors.

Multispecies ethnography focuses not only on various species but also on human relationships with both the natural environment and human-made infrastructure (Hamilton & Taylor, 2017). In a study by Atsuro Morita (2017), the Chao Phraya delta in Thailand, specifically in the provinces of Ayutthaya and Bangkok, was examined. The study aimed to understand the capacity of infrastructure to shape multispecies relations. To comprehend the relationship between the delta, floating rice, and other elements, Morita delved into previous studies on rice and water management. These studies introduced the research team to the actors involved and demonstrated how rice became a matter of care for farmers and the Thai state. Morita's ethnographic research explores the phenomenon of entanglement among infrastructure, farmers, and rice species. The study also emphasises that infrastructures are not merely engineering systems but involve multispecies components, forming what Morita refers to as the human-rice assemblage.

Lastly, in an article that examined non-living materiality in shaping urban conditions, Jakkrit Sangkhamanee (2021) explored the impact of cloudbursts on Bangkok's urban environment and atmosphere. Using more-than-human ethnography, the author argues that urbanity is a dynamic process in which heterogeneous socio-natural entities converge and collide, leading to the emergence of forms of urban sentience. The author examines how cloudbursts affect the city's atmosphere and how this impacts urban planning and development. The research suggests that understanding the impact of cloudbursts on Bangkok's urban environment can help make urban matters of concern visible and catalyse the city's missing masses.

The advantages and appropriateness of ethnography as a spatial method for studying urban sustainability include its ability to capture the complex social and ecological relationships between human and non-human actors in urban ecosystems, and its potential to challenge traditional assumptions about urban environments. However, this approach also has some limitations, including the potential for subjective interpretations of data and the difficulty of generalising findings to other contexts. The works on ethnographic approaches to spatial methods demonstrate how sensorial and affective approaches can provide valuable insights into the subjective and experiential dimensions of urban sustainability. Researchers can learn how different people perceive and experience urban sustainability by examining the sensory and emotional aspects of urban environments, which will help them develop policies and interventions that are more responsive to the needs and aspirations of urban residents. Overall, ethnography can provide valuable insights into the social and cultural dimensions of urban sustainability. By exploring the ways in which people interact with their urban environment, ethnography can reveal the lived experiences of sustainability and help inform policies and interventions that are better tailored to the needs of urban residents.

From the above review, we see that spatial methods can play a critical role in advancing SDG11 by providing precise spatial data, analytical tools, and participatory approaches that enable informed decision-making in urban planning and development. These methods enable the city to allocate resources more effectively, design resilient infrastructure, improve access to housing, services, and transportation, and engage communities in the process. We see spatial methodology as a useful tool for improving the quality of life for urban residents by leveraging the capabilities of GIS, spatial analysis, and ethnographic methodology to facilitate holistic urban planning, fostering well-designed, inclusive, and sustainable cities that align with the objectives of SDG11.

6.2 The Usefulness of Spatial Methods for Urban Sustainability Studies

As we have seen from the last section, spatial research methods offer a multifaceted understanding of urban sustainability by considering the spatial context, relationships, and dynamics within urban environments. Researchers can obtain insights into the interrelated social, economic, and environmental aspects of urban sustainability by combining spatial data and analysis techniques. This can result in more informed policymaking and interventions. This method of conducting spatial research is therefore regarded as transdisciplinary. As we use the term transdisciplinarity, we mean an approach to research, problem-solving, and knowledge creation that transcends the confines of individual academic disciplines. It entails active engagement with non-academic stakeholders, including practitioners, community members, policymakers, and industry professionals, in addition to interdisciplinary collaboration and interaction. The goal of transdisciplinarity in the application of spatial methods is to tackle difficult, real-world issues that elude complete comprehension or resolution within the parameters of a single discipline (Nicolescu, 2014; Ramadier, 2004).

In addition to understanding the issues and contexts from interdisciplinary perspectives, spatial methods can also play a vital role in fostering collaboration and allowing diverse partners and multiple stakeholders to work together to address urban issues. For instance, spatial methods, such as mapping and visualisation, offer a common language and a visual representation of urban issues. Spatial methods also facilitate communication and increase comprehension among diverse stakeholders, including policymakers, urban planners, community members, and researchers, by presenting data and analysis results visually. Stakeholders can be directly involved in data collection, analysis, and decision-making using spatial methods. Citizen science and participatory mapping allow community members to contribute with their knowledge, perspectives, and experiences. This participatory strategy in spatial methods can empower partners and encourage inclusiveness in decision-making among different stakeholders.

Finally, spatial methods allow for the integration and sharing of diverse datasets from multiple partners and stakeholders. By combining diverse data sources, such as satellite imagery, sensor data, and data generated by the community, stakeholders can gain a more comprehensive understanding of urban issues and work together to identify solutions. This also facilitates scenario planning and visualisation, allowing stakeholders to investigate various future scenarios and assess potential interventions. In many instances, as we will see in the section that follows, spatial methods foster collaborative planning and design processes, bringing together stakeholders with diverse expertise and points of view. By utilising spatial tools, stakeholders can design and evaluate urban interventions such as land use plans, transportation systems, and green infrastructure collaboratively, thereby fostering collective ownership and shared responsibility.

In short, as the case studies from Bangkok have shown, spatial methods can create a framework for multi-stakeholder collaboration by facilitating shared comprehension, participatory processes, data integration, scenario planning, and collaborative decision-making. By utilising spatial information and analysis, stakeholders can collaborate to address urban issues, thereby fostering inclusive and sustainable urban development.

6.3 Sensorial and Affective Spatial Methodology: An Ongoing Experiment from Ari Ecowalk

Bangkok has a vibrant and diverse urban environment. However, without a transdisciplinary research approach and numerous spatial tools, our hasty movements, and careless eyes as we go about our daily lives in the city may obscure such complexities and dynamism. In this final section, we would like to present a spatial methodological design process based on our experience in Bangkok to highlight the ongoing and persistent effort to come to terms with the appropriate, available methods for studying urban materiality, multispecies, and atmosphere. As the case will demonstrate, to make sense of the city's multiplicity, the methodological instrument must be tailored to the temporal and spatial situatedness of the urbanity, as the city is constantly evolving.

Ari Ecowalk is a slow-paced activity organised by volunteers to explore the city's neighbourhood to pay closer attention to the environment and the city's configuration. The name of the neighbourhood, Ari, means generosity and thoughtfulness. The purpose of the walk is to explore and connect with the generosity that urban space offers, as well as to reflect on our relationship with other humans and non-humans. The Ari neighbourhood is located right in the heart of Bangkok. There are numerous alleys and major roads that pass through its roughly 3-kilometre-square area. The area is a mix of low- and high-rise middle-class residences, government buildings and compounds, and corporate offices. Over the past few decades, the area has undergone rapid gentrification with the proliferation of trendy restaurants and cafes that attract daytime and night-time visitors from other parts of the city. Although the area is connected to BTS, Bangkok's skytrain system, footpaths are fragmented and limited, necessitating the use of motorcycle taxis, personal cars, taxis, and on-demand electric *tuk tuks* for most of the mobility.

Since its establishment in 2020, Ari Ecowalk has organised over 40 walks. Over the past few years, people from various backgrounds and demographics have actively participated in and contributed to the Ari Ecowalk. They are urbanists eager to learn more about their surroundings. They come from various backgrounds, but urban development is their main concern. Approximately 20–30% are permanent residents. The walk taught us that each brick, tree, and other non-human entity contains both the history of urban development and the natural history of the area. Space, or the environment, is intrinsically linked to our health. In the book *Last Child in the Wood*, Richard Louv (2005) argued that our physical, mental, and spiritual health are significantly affected by our connection to nature and the ability to comprehend one's surroundings could be termed 'sense of place'. Evidently, Bangkok's urbanisation and rush for mobility have made people lose touch with their sense of place. A poorly designed city plan exacerbated the problem by severing our ties with our own kind and non-human species. According to Chawla (2020), 'nature connection' could be rooted in an individual's mind, which in turn motivates constructive hope and actions. Therefore, if we establish strong relationships with the natural world, we protect our physical, mental, and spiritual health.

When Ecowalk was in its infancy, there were no established walk procedures in place. The walk concept was based on the notion that individuals perceive a specific place differently (Adams et al., 2017). Therefore, the project was intended to be experimental for both walk organisers and participants. Volunteers attempt to utilise as many methods as possible. Even though the walk lacks specific methods, it still contains significant points of interest. The essence of this project is knowledge and interconnectedness. The interaction between these two characteristics could promote environmental stewardship and heighten awareness of environmental injustice. The walk loosely employs three methods—observation, participation, and biophysical immersion—to gain knowledge and forge connections (Figs. 6.1 and 6.2).

Figs. 6.1 and 6.2 Participants in Ari Walk engaged in different sensorial activities. (Source: authors)

Observation is the foundation of knowledge inquiry. During the walk, the lack of strict rules permits people to pause and observe their surroundings closely. It also prevents the tendency to jump to premature conclusions. The more time participants devote to observing their surroundings, the more specific their descriptions become. Participation is the act of dwelling in the present. During the walk, participants are encouraged to participate in the environment by being there and noticing the interactions between themselves and the natural environment. Examples of participation in this case include hugging trees, attentively listening to bird songs, foraging for fruits and herbs, and feeling the wind and water temperature on our skin. Biophysical immersion reflects our thoughts and feelings as a part of Ecowalk. Since we tend to perceive the reality of places differently, the walk ensures that participants find their own ways to make sense of the day's experiences.

The walk revealed that these various methods could fall under the purview of sense of place and affective approaches to spatiality. Sense of place is the study of the relationships that people develop or experience with specific places and environments through place attachment and place meaning (Foote & Azaryahu, 2009; Foray, 2023). It also contributes to social-ecological sustainability (Masterson et al., 2017).

Each Ari Ecowalk procedure is highly dependent on its theme, orientation, and participants. Occasionally, the purpose of the walk was to learn about the urban smellscape and to try to understand what scents can reveal, such as consumption patterns, greenspace sizes, housing types, the number of public garbage bins, and the policy and implementation of garbage separation. During a different walk, the focus was on understanding how local species' coexistence with humans affects their behaviour and habitats. In addition to materiality and multispecies, the spatial method utilised during the walk may be highly abstract and expose the unique sensibilities of each participant. During one of the walks, the objective was to understand the city's activities, landscape, infrastructure, and atmosphere through smell. Other walks use sound mapping to evaluate the impact of human activity on local species.

Despite the variety of themes, orientations, and methods, the walk has a few essential components that have proven useful for organising and enhancing the experiences of participants. The walk leader will lead participants through the neighbourhood on foot. The start and finish locations are never identical and are never disclosed to participants unless they express concern. Every walk begins with a brief explanation of the ground rules, which are that there are no rules! Do what you enjoy doing. The walk leaders are neither the day's instructors nor any type of specialist. The facilitator of the walk would request that all participants use the objects in front of them as their guides, lessons, and instructions. One of the facilitators told the group, 'Respect is essential', without providing a precise definition of what and how to respect.

After a brief introduction, walk leaders typically encourage participants to familiarise themselves with their own senses, which are an excellent resource for making sense of our surroundings. For instance, the walk leader may ask questions that highlight the sensation of wind on the skin. In other locations, the walk may focus on the feathers and habitats of local birds and ask participants to examine them with binoculars. Alternately, the walk leaders may ask them to compare the aromas of a young and an older leaf from a tree along the path. Typically, the walk co-creates these experiences with nature and urban infrastructure; therefore, it is dependent on the materials and nature available on any given day. While acclimating to external stimuli, the walk employs a series of close observations, such as sensory experiences, critical questions, and narratives of urban wildlife, to unlock the 'desire to absorb'. These tools may include understory plant life, landscape typology, a bird's nest, fruits dangling from a wall, tree symptoms or shapes and forms, balcony gardens, etc.

People become exhausted after at least 2 hours of urban heat and a long walk. It is now time for solitude. The walk leader will lead them to a more pristine and tranquil environment and request that they be alone and refrain from conversing with anyone else. Participants will sit quietly in the shade or by a pond. The walk leader will occasionally introduce common nature games such as sound mapping, mood by colour, a sit spot, and poetry writing. This would afford them the opportunity to develop a close relationship with nature and with themselves.

After participants have had a complete encounter with urban nature, the walk leader will gather them in a circle and begin asking them questions. This is the opportunity for participants to reflect on their walking experience. Typically, three questions will be asked; 'what did you notice?', 'What are your impressions of the experience?', and 'What is the environment?' The questions will be asked sequentially. On post-it notes, participants will be required to write their own reflections. The facilitator will then ask them to share their writing with the group. Everyone has an equal opportunity to speak. After everyone has shared their thoughts, the session will typically end without a conclusion.

Overall, Ari Ecowalk has contributed to the study, exposure, comprehension, and promotion of the urban's sustainability with more than 120 participants to date. It provides people with direct experiences with the surrounding natural environment through place-based learning. Through ecoservices and their connection to nature, people are made aware of the rights to life of other beings and their contributions to our well-being. Additionally, the walk sparks a discussion about environmental justice. It influences anthropocentrism-to-ecocentrism cultural transitions in urban settings.

6.4 Conclusion

In this chapter, we have examined the hybrid use of spatial methods in transdisciplinary urban sustainability studies on and from Bangkok. We argued that spatial methods can be used to examine the complex relationships between social, economic, and environmental factors in urban areas. Based on perspectives from the Thailand capital, this chapter highlighted the hybrid applicability and suitability of recent research on spatial methods for transdisciplinary studies of urban sustainability in relation to a variety of stakeholders, contexts, and issues. In this contribution, we demonstrated from a Bangkokian perspective that distinct spatial methods are uniquely suited to the study of distinct issues, their application in distinct contexts, and their use by distinct partners. Due to the complexity and dynamism of urban conditions, spatial methods must be refined and combined with other research methods, with the participation of transdisciplinary partners and stakeholders, to reflect the conditions of urban issues and sustainability.

We have highlighted that spatial methods, particularly GIS, spatial analysis, and ethnography, play a crucial role in understanding and enhancing urban resilience. Urban resilience, as we suggested, refers to the ability of urban systems to withstand, adapt to, and recover from various challenges, including natural disasters, climate change, and socio-economic disruptions. GIS and spatial analysis are pivotal in this regard as they provide a framework for capturing, storing, analysing, and visualising geospatial data. This capability is essential in urban planning and disaster management, where understanding the spatial distribution of resources, risks, and population demographics is critical. As we have shown, GIS can be used to model the impact of natural disasters like floods on urbanity, helping in risk assessment and mitigation planning. Spatial analysis further aids in identifying patterns and trends in urban growth and resource distribution, enabling the identification of vulnerable areas and the optimisation of resource allocation.

Ethnography, on the other hand, offers a qualitative perspective by exploring the lived experiences of urban residents. It provides insights into how communities interact with their environment and respond to challenges, contributing to a more nuanced understanding of urban resilience. This qualitative approach complements the quantitative data obtained from GIS and spatial analysis, leading to a more holistic view. When combined, these methods can inform and enhance collaboration among various partners in research and policymaking. Policymakers can leverage the insights obtained from GIS and spatial analysis to design evidence-based strategies, while ethnographic studies can ensure these strategies are grounded in the realities of local communities. This integrated approach fosters a more inclusive and effective decision-making process, accommodating diverse perspectives and needs. Consequently, the combination of spatial methods and ethnography can significantly contribute to building more resilient urban systems that are better equipped to face future challenges.

In terms of collaboration, these spatial methods serve as a common language and framework that can bridge the gap between researchers, policymakers, and communities. As we have shown, GIS and spatial analysis provide clear, visual representations of data that are easily understandable and can be used to communicate complex urban issues to a broad audience. This facilitates more informed decision-making and public engagement. Ethnography, on the other hand, ensures that the voices and perspectives of local communities are included in the conversation, promoting more inclusive and participatory approaches to urban resilience. Together, these methods can foster a more integrated, multidisciplinary approach to urban resilience, combining scientific analysis with community insights to develop strategies that are both effective and socially responsive.

Our experience with Bangkok's Ari Ecowalks also illuminates how diverse orientations, stakeholders, temporality, and contexts influence the applicability of spatial methods in urban sustainability studies. The sensorial and affective spatial methodology put forth by Ari Ecowalk embodies a pioneering spirit in urban exploration and engagement especially in Bangkok. It underscores the intricate tapestry of urban life, where each thread—be it the quiet buzz of the natural world or the stark imprint of human development—contributes to the richness of our urban experiences. The methodology underscores the importance of multi-sensory engagement with our surroundings, encouraging participants to truly inhabit the moment and engage with the environment in a meaningful way. By fostering this deep, personal connection with urban spaces, Ari Ecowalk challenges the prevailing notions of urban anonymity and detachment. It illuminates the oft-overlooked nooks of urbanity, invigorating a sense of communal stewardship and ecological mindfulness. This initiative not only enriches the participants' understanding of their environment but also incubates a collective consciousness and respect for the delicate balance between urban development and natural harmony. The essence of Ari Ecowalk lies not just in the steps taken but, in the senses, awakened and the connections forged, paving the way for a more resilient, aware, and harmonious coexistence within our urban ecosystems.

References

Adams, J. D., Greenwood, D. A., Thomashow, M., & Russ, A. (2017). Sense of place. In A. Russ & M. E. Krasny (Eds.), *Urban environmental education review* (pp. 68–75). Cornell University Press.

Ahamed-Broadhurst, K. E. (2018). *Understanding canals in Bangkok using historic maps and GIS* [Doctoral dissertation, Harvard University]. Digital access to scholarship at Harvard. https://dash.harvard.edu/handle/1/37736775

Amin, A. (2014). Lively infrastructure. *Theory, Culture & Society, 31*(7–8), 137–161.

Anderson, B., & Holden, A. (2008). Affective urbanism and the event of hope. *Space and Culture, 11*(2), 142–159.

Askew, M. (1999). Community-building among the Bangkok middle class: Ethnographic perspectives on group identity and problem-solving in a suburban housing estate. *South East Asia Research, 7*(1), 93–120.

Brenner, N., Madden, D. J., & Wachsmuth, D. (2011). Assemblage urbanism and the challenges of critical urban theory. *City, 15*(2), 225–240.

Cadieux, K. V. (2008). Political ecology of exurban "lifestyle" landscapes at Christchurch's contested urban fence. *Urban Forestry & Urban Greening, 7*(3), 183–194.

Chawla, L. (2020). Childhood nature connection and constructive hope: A review of research on connecting with nature and coping with environmental loss. *People and Nature, 2*(3), 619–642.

Checker, M. (2011). Wiped out by the "Greenwave": Environmental gentrification and the paradoxical politics of urban sustainability. *City & Society, 23*(2), 210–229.

Clough, P. T., & Hallry, J. (Eds.). (2007). *The affective turn: Theorizing the social*. Duke University Press.

Cohen, E. (2012). Flooded: An auto-ethnography of the 2011 Bangkok flood. *ASEAS-Austrian Journal of South-East Asian Studies, 5*(2), 316–334.

Di Chiro, G. (2018). Canaries in the anthropocene: Storytelling as degentrification in urban community sustainability. *Journal of Environmental Studies and Sciences, 8*, 526–538.

Farias, I. (2011). The politics of urban assemblages. *City, 15*(3–4), 365–374.

Foray, R. (2023). Affective landscapes: Capturing emotions in place. In S. A. Lovell, S. E. Coen, & M. W. Rosenberg (Eds.), *The Routledge handbook of methodologies in human geography* (pp. 109–122). Routledge.

Foote, K. E., & Azaryahu, M. (2009). Sense of place. *International Encyclopedia of Human Geography*, 96–100.

Giordano, A. (2023). Towards Interdisciplinarity: The relationship between GIS/GIScience/cartography and human geography. In S. A. Lovell, S. E. Coen, & M. W. Rosenberg (Eds.), *The Routledge handbook of methodologies in human geography* (pp. 47–60). Routledge.

Gullion, J. S. (2018). *Diffractive ethnography: Social sciences and the ontological turn*. Routledge.

Hamilton, L., & Taylor, N. (2017). *Ethnography after humanism: Power, politics and method in multi-species research*. Palgrave Macmillan.

Hawken, S., Rahmat, H., Sepasgozar, S. M. E., & Zhang, K. (2021). The SDGs, ecosystem services and cities: A network analysis of current research innovation for implementing urban sustainability. *Sustainability, 13*(24), 1–36.

Holbraad, M., & Pedersen, M. A. (2017). Introduction: The ontological turn in anthropology. In M. Holbraad & M. A. Pedersen (Eds.), *The ontological turn: An anthropological exposition* (pp. 1–29). Cambridge University Press.

Huang, G., & London, J. K. (2016). Mapping in and out of "messes": An adaptive, participatory, and transdisciplinary approach to assessing cumulative environmental justice impacts. *Landscape and Urban Planning, 154*, 57–67.

Iamtrakul, P., Srivanit, M., & Klaylee, J. (2017). Resilience in urban transport towards hybrid canal-rail connectivity linking Bangkok's canal networks to mass rapid transit lines. *International Journal of Building, Urban, Interior and Landscape Technology (BUILT), 10*, 27–42.

Kamal, N., Imran, M., & Tripati, N. K. (2017). Greening the urban environment using geospatial techniques, a case study of Bangkok, Thailand. *Procedia Environmental Sciences, 37*, 141–152.

Koh, R., Babel, M. S., Shinde, V. R., & Mendoza, G. (2022). Towards climate resilient municipal water supply in Bangkok: A collaborative risk informed analysis. *Climate Risk Management, 35*(100), 406.

Laeni, N., van den Brink, M., & Arts, J. (2019). Is Bangkok becoming more resilient to flooding? A framing analysis of Bangkok's flood resilience policy combining insights from both insiders and outsiders. *Cities, 90*, 157–167.

Logan, J. R. (2012). Making a place for space: Spatial thinking in social science. *The Annual Review of Sociology, 38*, 507–524.

Louv, R. (2005). *Last child in the woods: Saving our children from nature-deficit disorder*. Atlantic Books.

Malcolm, N. (2002). *The integration of remote sensing and GIS to facilitate sustainable urban environmental management: The case of Bangkok, Thailand* [Master's thesis, University of Waterloo]. UWSpace. https://uwspace.uwaterloo.ca/handle/10012/996

Masterson, V. A., Stedman, R. C., Enqvist, J., Tengö, M., Giusti, M., Wahl, D., & Svedin, U. (2017). The contribution of sense of place to social-ecological systems research: A review and research agenda. *Ecology and Society, 22*(1).

McFarlane, C. (2011a). Assemblage and critical urbanism. *City, 15*(2), 204–224.

McFarlane, C. (2011b). The city as assemblage: Dwelling and urban space. *Environment and Planning D: Society and Space, 29*(4), 649–671.

Monprapussorn, S., & Phoung Ha, L. (2021). Integrated analysis of climate, land use and water for resilience urban megacities: A case study of Thailand and Viet Nam. *APN Science Bulletin, 11*(1), 74–80.

Morita, A. (2017). Multispecies infrastructure: Infrastructural inversion and involutionary entanglements in the Chao Phraya Delta, Thailand. *Ethnos, 82*(4), 738–757.

Neupane, S., Ghimire, U., Shrestha, S., & Mohanasundaram, S. (2022). Mapping groundwater resilience to climate change and human development in Bangkok and its vicinity, Thailand. *APN Science Bulletin, 13*(1), 163–198.

Nicolescu, B. (2014). Methodology of transdisciplinarity. *The Journal of New Paradigm Research, 70*(3–4), 186–199.

Ramadier, T. (2004). Transdisciplinarity and its challenges: The case of urban studies. *Futures, 36*(4), 423–439.

Ratanawaraha, A. (2016). Inequality, fragility, and resilience in Bangkok. In F. Mancini & A. Ó. Súilleabháin (Eds.), *Building resilience in cities under stress* (pp. 4–12). International Peace Institute.

Pedro, J., Silva, C., & Pinheiro, M. D. (2019). Integrating GIS spatial dimension into BREEAM communities' sustainability assessment to support urban planning policies, Lisbon case study. *Land Use Policy, 83*, 424–434.

Pinfold, N., & Mokhele, M. (2019). Mapping memories through geographic information system: The case of St mark's transdisciplinary service-learning project in District Six, Cape Town. *Alternation [Special issue], 29*, 223–244.

Posuk, S., Kajita, Y., & Petchsasithon, A. (2018). Comparative analysis of city planning and land use change in Bangkok, Thailand, by using remote sensing and GIS. *MATEC Web of Conferences, 192*, 02064.

Rongwiriyaphanich, S. (2012, May). Effects of land policy on hybrid rural-urban development patterns and resilience: A case study of the territorial development in the Bangkok Metropolitan Region. In *Regional Studies Association European Conference, Delft, Netherlands* (Vol. 15).

Sands, K. L. (2009). Shared spaces on the street: A multispecies ethnography of ex-racing greyhound street collections in South Wales, UK. *Leisure Studies, 38*(3), 367–380.

Sangkhamanee, J. (2021). Bangkok precipitated: Cloudbursts, sentient urbanity, and emergent atmospheres. *East Asian Science, Technology and Society: An International Journal, 15*(2), 153–172.

Savvides, N. (2013). Living with dogs: Alternative animal practices in Bangkok, Thailand. *Animal Studies Journal, 2*(2), 28–50.

Sheller, M., & Urry, J. (2006). The new mobilities paradigm. *Environment and Planning A: Economy and Space, 38*(2), 207–226.

Sintusingha, S. (2011). Bangkok's urban evolution: Challenges and opportunities for urban sustainability. In A. Sorensen & J. Okata (Eds.), *Urban form, governance, and sustainability* (pp. 133–161). Springer.

Sitko, P., & Almuhktar, A. (2017). Understanding urban disaster resilience through a morphological approach: A case study of settlement upgrading and flood response in Bangkok. *Journal of Research in Architecture and Planning, 23*(2), 22–30.

Sivado, A. (2015). The shape of things to come? Reflections on the ontological turn in anthropology. *Philosophy of the Social Sciences, 45*(1), 83–99.

Sivado, A. (2020). Ways to be understood: The ontological turn and interpretive social science. *Philosophy of the Social Sciences, 50*(6), 565–585.

Sopranzetti, C. (2018). Towards an ethnography of urban circulation. In S. Low (Ed.), *The Routledge handbook of anthropology and the city* (pp. 113–125). Routledge.

Thaiutsa, B., Puangchit, L., Kjelgren, R., & Arunpraparut, W. (2008). Urban green space, street tree and heritage large tree assessment in Bangkok, Thailand. *Urban Forestry & Urban Greening, 7*(3), 219–229.

Thanvisitthpon, N., Shrestha, S., & Pal, I. (2018). Urban flooding and climate change: A case study of Bangkok, Thailand. *Environment and Urbanization ASIA, 9*(1), 86–100.

Thibaud, J.-P. (2015). The backstage of urban ambiences: When atmospheres pervade everyday experience. *Emotion, Space and Society, 15*, 39–46.

Thrift, N. (2014). The 'sentience' city and what it may portend. *Big Data & Society, 1*(1), 1–21.

Timur, S., & Getz, D. (2008). A network perspective on managing stakeholders for sustainable urban tourism. *International Journal of Contemporary Hospitality Management, 20*(4), 445–461.

Turner, B. (2011). Embodied connections: Sustainability, food systems, and community gardens. *Local Environment, 16*(6), 509–522.

Wei, J., Qian, J., Tao, Y., Hu, F., & Ou, W. (2018). Evaluating spatial priority of urban green infrastructure for urban sustainability in areas of rapid urbanization: A case study of Pukou in China. *Sustainability, 10*(2), 327.

Yamashita, A. (2017). Bangkok metropolitan area: Geospatial analysis of metropolises. In Y. Murayama, C. Kamusoko, A. Yamashita, & R. C. Estoque (Eds.), *Urban development in Asia and Africa* (pp. 151–169). Springer.

Jakkrit Sangkhamanee is an Associate Professor of Anthropology at the Chulalongkorn University Faculty of Political Science in Bangkok. His work focuses on Science and Technology Studies, specifically hydrological engineering projects related to the Thai state formation and climate politics. His latest publications include 'An Assemblage of Thai Water Engineering: The Royal Irrigation Department's Museum for Heavy Engineering as a Parliament of Things' (*Engaging Science, Technology and Society*, vol. 3, 2017), 'Infrastructure in the Making: The Chao Phraya Dam and the Dance of Agency' (*TRaNS: Trans-Regional and -National Studies of Southeast Asia*, 6(1), 2018) and 'Bangkok Precipitated: Cloudbursts, Sentient Urbanity, and Emergent Atmospheres' (East Asian Science, Technology and Society: An International Journal (EASTS), 15(2), 2021).

Piyathep Tanmahasmut is an architectural and design anthropologist who worked with urban planners and architects in Bangkok's CBD and cultural areas before starting to cooperate with the Bangkok Teak Research by using ethnographic methodology to develop urban solutions for affordable housing, multi-generational homes, and mixed-use spaces. Tanmahasmut is also an MA student at Chulalongkorn University, where he pursues his interests in urban settlements, Anthropocene, and urban assemblage. His research focuses on shophouse residents in Chinese Thai neighbourhoods facing gentrification.

Poramin Watnakornbancha is a marketer and a naturalist. He has been working on the Ecowalk initiative, a place-based learning program aiming to advocate human-nature connection in the realm of biodiversity, since 2019. His works focus on collecting Ari-Pradiphat's biodiversity data, addressing environmental injustice and promoting benefits of urban nature ecosystem services. He is keen on utilising data derived from Ecowalk devising Bangkok's green policies. Amidst Climate crisis, the cities are his chosen battlefield for change.

Open Access This chapter is licensed under the terms of the Creative Commons Attribution 4.0 International License (http://creativecommons.org/licenses/by/4.0/), which permits use, sharing, adaptation, distribution and reproduction in any medium or format, as long as you give appropriate credit to the original author(s) and the source, provide a link to the Creative Commons license and indicate if changes were made.

The images or other third party material in this chapter are included in the chapter's Creative Commons license, unless indicated otherwise in a credit line to the material. If material is not included in the chapter's Creative Commons license and your intended use is not permitted by statutory regulation or exceeds the permitted use, you will need to obtain permission directly from the copyright holder.

Implementing a Transdisciplinary Approach in Flood Risk Management: Insights from Tangerang, Indonesia

Raldi Hendro Koestoer and Budi Heru Santosa

7.1 A Path Towards Resilient Flood Management: Transdisciplinarity, Systemic, Community-Based and Spatial

Flood is a complex problem that affects many aspects of society and the environment. Therefore, it requires a complex and holistic solution that considers multiple factors and perspectives (Thaler et al., 2021). One widely accepted solution is disaster risk reduction (DRR), which aims to reduce the vulnerability and exposure of people and assets to flood hazards. However, DRR cannot be achieved by a single academic discipline or sector alone; it demands inter- and transdisciplinary collaboration among various stakeholders (Takara, 2018). In this study, transdisciplinarity can be described as a comprehensive research method that tackles societal concerns and scientific challenges by bringing together academic and non-academic participants and incorporating various knowledge domains into the process. The transdisciplinary approach is a promising way to foster such collaboration, as it enables the integration of different types of knowledge and the co-creating innovative and cooperative solutions based on trust, understanding and mutual reliance (Almoradie et al., 2020). Moreover, the transdisciplinary approach can provide an efficient communication tool and thinking system to help formulate flood resilience strategies (Fratini et al., 2012).

In recent years, there has been a substantial shift in the approach to flood risk management, with a strong emphasis on promoting community resilience as a practical strategy to address this pressing threat (Bertilsson et al., 2019). The concept of resilience, defined in various ways by scholars, essentially refers to a system's capacity to withstand and recover from shocks, ultimately returning to its initial state or adjusting to the changes induced by the disturbance (IPCC, 2022). Achieving resilience involves restoring the system to its initial state through engineering interventions and enhancing the well-being of society, the economy and public health, thereby fostering community cohesion (Vardoulakis et al., 2022). In addition to understanding flood resilience, it is essential to grasp the factors influencing river basin flood susceptibility to enable practical efforts to reduce such susceptibility (Santosa et al., 2022). Flood-related issues have traditionally been addressed by specialists from various disciplines without systematic integration, leading to suboptimal solutions. However, in recent years a growing body of research has

R. H. Koestoer (✉)
School of Environmental Science,
Universitas Indonesia, Jakarta, Indonesia
e-mail: raldy.hk@ui.ac.id

B. H. Santosa
National Research and Innovation Agency (BRIN), Jakarta, Indonesia

advocated for a more interdisciplinary approach that integrates disciplinary knowledge, problem formulation, methods and data to tackle flood-related challenges (Tate et al., 2021). This integrated approach has proven to be more successful in providing better solutions than previous methods (Sadiq et al., 2019).

Scholars have developed various methods to assess river basin flood susceptibility, including statistical, hydrodynamic and spatial approaches (Narendra et al., 2023). With the advancement of remote sensing and GIS technologies, geospatial methods have become one of the practical approaches to generating river basin susceptibility maps (Bui et al., 2019). Numerous academic works have explored using geospatial analyses based on multiple criteria to assess flood susceptibility. Some studies have attempted to address this limitation by combining a geospatial approach with a multi-criteria decision-making analysis (Feloni et al., 2020; Nsangou et al., 2022). Several studies have integrated the random forest algorithm into GIS approaches (Li et al., 2021) or employed the fuzzy method (Ekmekcioğlu et al., 2021). Furthermore, other studies have applied a transdisciplinary approach involving community participation in mapping physical flood vulnerability (Minucci et al., 2020). The utilisation of participatory flood risk mapping enables communities to identify potential flood countermeasures, which they can implement to mitigate losses and safeguard their livelihoods (Samaddar et al., 2022). This situation aligns with the goal of enhancing flood resilience, since achieving it would empower flood-affected communities to quickly return to their normal daily activities following a flood event.

When evaluating community flood resilience, it is of utmost importance to incorporate perspectives from field practitioners and stakeholders who have first-hand experience with flooding. The assessment of resilience from a subjective standpoint assumes a pivotal role, involving mental and emotional self-judgments of individuals about their ability to cope with risks while considering community views (Jones & Tanner, 2017). The subjective approach does not prioritise expert opinions as the primary assessment of community resilience in facing disasters (Bottazzi et al., 2018). Instead, it focuses on individual opinions as a more appropriate way to understand individual conditions in the context of flood resilience. The subjective resilience method involves gathering the opinions of community members affected by floods to assess their specific conditions using various data acquisition techniques employed within the context of the subjective resilience approach, aiming to triangulate information obtained from diverse sources (Santosa et al., 2023). These techniques include questionnaire surveys, in-depth interviews and focus group discussions (Sun et al., 2022). Using this method, researchers comprehensively understand community members' perspectives and experiences, leading to a complete evaluation of flood resilience at the community level. Such comprehensive and participatory approaches enable a deeper insight into the community's ability to cope with and adapt to flood risks, thus enhancing the effectiveness of resilience-building strategies and disaster management efforts.

In flood-prone regions, improved community flood resilience is essential to sustain habitation and ensure residents' well-being. Therefore, this study aims to explore implementing a transdisciplinary approach in flood risk management based on field practises. To this end, we employed a participatory geospatial approach to assess river basin flood susceptibility, utilising land cover data spanning from 2001 to 2021. In addition, we conducted an examination of flood resilience within the community in our study area, soliciting subjective viewpoints from households that have been affected by flooding and from local community leaders. A comprehensive understanding of river basin flood susceptibility and community flood resilience could greatly assist city planners and local authorities in addressing flood disaster risks, including those affecting the local community. Consequently, the community can take proactive measures to mitigate potential flood situations in the future. Notably, successful implementation of such integration could be crucial in achieving Sustainable Development Goal 11, which aims to 'Make cities and human settlements inclusive, safe, resilient and sustainable', with a particular emphasis on Targets 11.5 and 11.B. These targets underscore the importance of significantly reducing the number of deaths and people affected, substantially decreasing direct economic losses, enhancing urban resilience to disasters like floods and promoting participatory approaches in urban planning and management.

7.2 Making Perceptions Spatially Tangible: Bringing the Spatial Dimension of Risk to the Fore

Disaster risk management planning, including flood risk management, can be strengthened by applying a transdisciplinary approach, facilitating productive collaboration between academic and local authorities and involving non-academic stakeholders (Marchezini, 2020). Stakeholders can engage in dialogue from diverse perspectives to collectively achieve knowledge and solutions in flood risk management (Almoradie et al., 2020). This inclusive approach can enhance disaster preparedness and response strategies to achieve a more resilient community in the face of flooding. Additionally, the transdisciplinary approach with early stakeholder involvement is beneficial in mitigating and eliminating controversial narratives that may hinder the study process (Thaler et al., 2021). The interdependency between upstream and downstream areas must be identified in a river basin to understand the flood risks faced in the downstream areas. One specific spatial method that can be applied is the geospatial method, which has been utilised in the *Kali* Ledug River basin in Tangerang, Indonesia (Santosa et al., 2022). Using geospatial methods helps identify areas with high flood susceptibility, enabling communities to prepare for floods and achieve adequate flood resilience. Effective communication of flood susceptibility to all stakeholders, including communities in the affected areas, neighbourhood leaders and local authorities, is crucial to achieving adequate flood risk perception. Risk perception is a crucial factor influencing the community's preparedness to face flood occurrences through implementing mitigation measures. Perception is a significant aspect that drives the initiation of efforts to reduce the impact of floods, whether at the household or societal level. The dissemination of a flood susceptibility map to the communities affected by floods will increase their awareness of flood events and should ultimately be expected to enhance community flood resilience.

7.2.1 Study Areas

This study was conducted in the *Kali* Ledug River Basin, situated in the urban area of Tangerang City, Indonesia, to the west of the capital city, Jakarta (Fig. 7.1). The river basin covers a relatively small area of approximately 14.45 km^2 and is characterised by a lowland basin with elevations ranging from 5 to 34.84 m above sea level, featuring a relatively uniform topography. The *Kali* Ledug is the primary watercourse within this basin, flowing and discharging into the Cirarab River. The rapid growth of the *Kali* Ledug River basin can be attributed to the urban development of the national capital, Jakarta.

The Tangerang City Government has strategically planned the development of the Periuk District area to encompass trading and service zones, medium to high-density housing and an environmentally sustainable integrated industrial area in their 2012–2032 Spatial Plan. Rapid physical development has consequently been undertaken to accommodate the region's economy, housing, industry and infrastructure needs. However, this significant development has increased flood discharge in the downstream area, resulting in periodic flooding, especially during the rainy season. Three sub-districts (SD1, SD2, SD3) have therefore become flood locations with regular occurrence since the late 1990s.

7.2.2 River Basin Flood Susceptibility Analysis

We employed several spatial data including slope level, altitude, distance from river, downstream

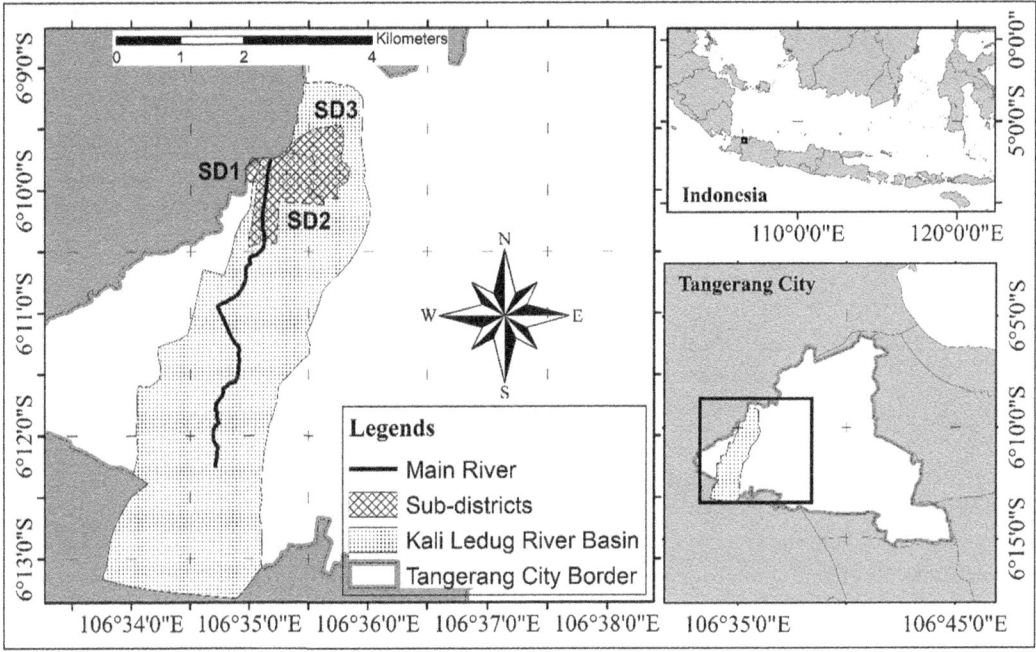

Fig. 7.1 *Kali* Ledug River Basin study area. (Source: created by authors)

distance, Topographic Wetness Index (TWI), soil type, land cover and precipitation in the spatial analyses to acquire pertinent information regarding river basin flood susceptibility. The spatial data was pre-processed with GIS software and transformed into raster data for spatial analysis. The thematic maps were then partitioned into distinct classes and evaluated to classify parameters based on their impact levels (Nsangou et al., 2022). Reclassification was utilised to assign a value to each raster in the spatial data using a scale of five levels, ranging from 1 (very low) to 5 (very high).

This study utilised spatial data from diverse sources with varying formats and resolutions. The Geospatial Information Agency (BIG) of Indonesia supplied the National Digital Elevation Model (DEMNAS) with a 0.27-arcsecond (8.3 m) resolution, crucial for delineating river basin boundaries and deriving altitude, slope and Topographic Wetness Index (TWI). The Landsat 7 ETM+ and Landsat 8 OLI optical bands were employed to acquire land cover data. River lines were extracted from satellite images and cross-validated with DEMNAS and BIG's thematic river data. The dataset facilitated the computation of distances from the river and downstream areas. Soil-type information originated from the Government of Tangerang City, while daily maximum precipitation data was sourced from the Indonesian Agency for Meteorology, Climatology and Geophysics (BMKG). Spatial variables were projected to the WGS 1984 UTM Zone 48S coordinate system and transformed to a raster format for comprehensive analysis. Numerous scholars have extensively investigated the influence of various types of spatial data on flood susceptibility, encompassing factors such as slope, altitude, river and downstream distances, topographic wetness index, soil type and land use.

This study employed a multi-criteria spatial analysis using GIS software to assess flood susceptibility in a specific geographical region. The analysis incorporated higher-value factors representing heightened flood susceptibility (Mukherjee & Singh, 2020). A flood susceptibility model was generated by applying weighted overlay analysis on raster cells, categorising susceptibility levels on a scale of 1 (very low susceptibility) to 5 (very high susceptibility). The model was subsequently validated using flood inundation data from a flood event in February 2020. Communities impacted by the flood were closely involved in the data collection, offering vital insights and on-the-ground assessments of the flood's effects. This approach ensured a high degree of concordance between the modelling outcomes and actual flood data, enhancing the model's reliability. This comprehensive empirical field study was conducted between 1 March and 30 June 2021.

Subsequently, the validated model was utilised to evaluate flood susceptibility dynamics over time by analysing changes in land cover from 2001 to 2021. This investigation resulted in the generation of five flood susceptibility maps representing 2001, 2006, 2011, 2016 and 2021. The model was applied to precipitation data with varying recurrence intervals from 10 to 100 years to forecast changes in flood susceptibility. As a result, a set of flood susceptibility maps corresponding to different precipitation return intervals was produced. Moreover, the flood susceptibility maps will be shared with the flood-affected community in the final phase to solicit feedback based on their experiences and observations regarding floods. This participatory approach aims to enhance the accuracy and relevance of the flood susceptibility model while fostering community engagement in flood risk management strategies.

7.2.3 Community Flood Resilience Analysis: Listening to Those Who 'Are in the Same Boat'

In the planning of flood risk management for communities, there has been a growing recognition of the value of adopting a transdisciplinary approach facilitated by a participatory methodology. This approach has gained prominence as a powerful strategy for the comprehensive integration of diverse perspectives from various segments of society (Räsänen et al., 2020). Several methodologies have been utilised to bolster community resilience against floods, including the approach by Nguyen and James (2013), which focuses on evaluating the community's capacity to mitigate flood hazards, diverging from conventional expert-dependent assessments. In this context, subjective resilience closely correlates with the community's introspective evaluation of their capacities to tackle flood-related risks (Jones & Tanner, 2017). Nevertheless, it is essential to note that ongoing efforts are being dedicated to further refining the measurement of this subjective community resilience (Clare et al., 2017).

The multifaceted nature of factors influencing community flood resilience encompasses many intricate elements. Mondal et al. (2021) and Snel et al. (2021) found that residents' expectations and perceptions of stakeholder responsibilities have substantial implications for effective flood management. Household economics plays a significant role in shaping resilience dynamics. Findings from a study conducted by Houston et al. (2020) in Scotland revealed that the immediate impacts of floods exhibit relatively equitable distribution, yet their influence diminishes over time. A sole factor does not drive the modulation of resilience; instead, it involves a complex interplay of socio-economic determinants that govern losses triggered by floods. Mashi et al. (2020) identified a range of scenarios where collaborative efforts between local government entities overseeing physical planning and waste management, alongside the limitations of emergency response mechanisms, compel community residents to engage in flood mitigation measures at the household level, even if this incurs economic burdens. Such investigation underscores the pivotal role of sociocultural factors and income in shaping flood adaptation endeavours.

This study applied the composite indicator method to accommodate the previously discussed factors. The method measured flood resilience using several indicators through identification and aggregation, resulting in a single measure representing a complex social issue, facilitating a comprehensive understanding of the system under investigation (Baptista, 2014). In addition, to determine the weight of each criterion that forms flood resilience, the analytical hierarchy process (AHP) is used (Saaty, 2008), which uses expert judgement through pairwise comparison of each indicator (Beccari, 2016). The AHP structures complex decision problems into a hierarchical system by applying a pairwise comparison matrix for indicators, using expert judgement to assign weights to them (Saaty, 2008).

In this study, household opinion data was obtained through questionnaires, while the opinions of neighbourhood leaders and local authorities were collected through in-depth interviews. Seven variables, encompassing 28 indicators, were employed to assess flood resilience: social conditions, economic conditions, home environment, communication and information, social capital, institutions and flood risk perception. The questionnaires were completed by flood-affected households across three sub-districts, totalling 354 households selected through proportional random sampling. Subsequently, the data from the questionnaire survey underwent meticulous processing and statistical analysis, employing predefined methods to assess its validity and reliability; the validity assessment aimed to ensure that the questionnaire served as a measuring instrument with robust capability. Meanwhile, reliability testing was employed to gauge the consistency of the questionnaire's outcomes upon repeated application. The reliability level was measured using the established Cronbach Alpha value.

In-depth interviews were conducted with ten neighbourhood leaders and heads of local authorities to enrich the understanding of flood events within the study area. The resulting interview data was transcribed into text and then subjected to thematic analysis, a qualitative method that categorises data into thematic groups representing the core ideas of a conversation. This approach aims to identify recurring patterns and relationships linking themes, ultimately forming a conceptual framework that elucidates the research subject (Braun & Clarke, 2006). The insights emerging from this thematic analysis were subsequently employed to validate the questionnaire survey results based on data from flood-affected households.

7.3 Results and Discussion

7.3.1 Participatory Spatial River Basin Flood Susceptibility

After determining the weights for each variable, a spatial analysis was conducted using a multi-criteria evaluation approach, with specific weights assigned to each criterion to analyse flood susceptibility factors. The weighted overlay method in the ArcGIS software facilitated quick and accurate spatial analysis. Validation was iteratively performed by adjusting the weight values of variables and observing the shape of areas with high susceptibility levels, using flood inundation data from February 2020 to obtain a valid model generating a flood susceptibility map (Fig. 7.2a). The validated model revealed varying impacts of each criterion on flood susceptibility, with altitude (41%) and distance from river (20%) having the most substantial influence. The additional considered factors were slope level (10%), distance downstream (8%), topographic wetness index (6%), land use land cover (7%) and precipitation (7%). Soil type had the most negligible impact (1%) on flood susceptibility. In the validated model, areas with a high degree of susceptibility cover 163.91 hectares (11.36% of the river basin area), while moderately susceptible areas cover 618.22 hectares (42.84%). Furthermore, low susceptible areas encompass 655.18 hectares (45.40%), and very low susceptible areas account for 5.70 hectares (0.40%). Based on the dominant variable weights affecting flood susceptibility, namely, altitude with 41% and distance from river with 20%, it can be generally concluded that areas with low altitudes (0–8 m) and a distance of less than 100 m from the river will have high susceptibility to flooding.

After constructing the validated model, an assessment of flood susceptibility in the *Kali* Ledug River basin was conducted. This evaluation utilised time series data from 2001 to 2021, including landcover information in a 5-year time interval. The evaluation results indicate that areas with high flood susceptibility have shown an almost linear increase from 2001 to 2021. During

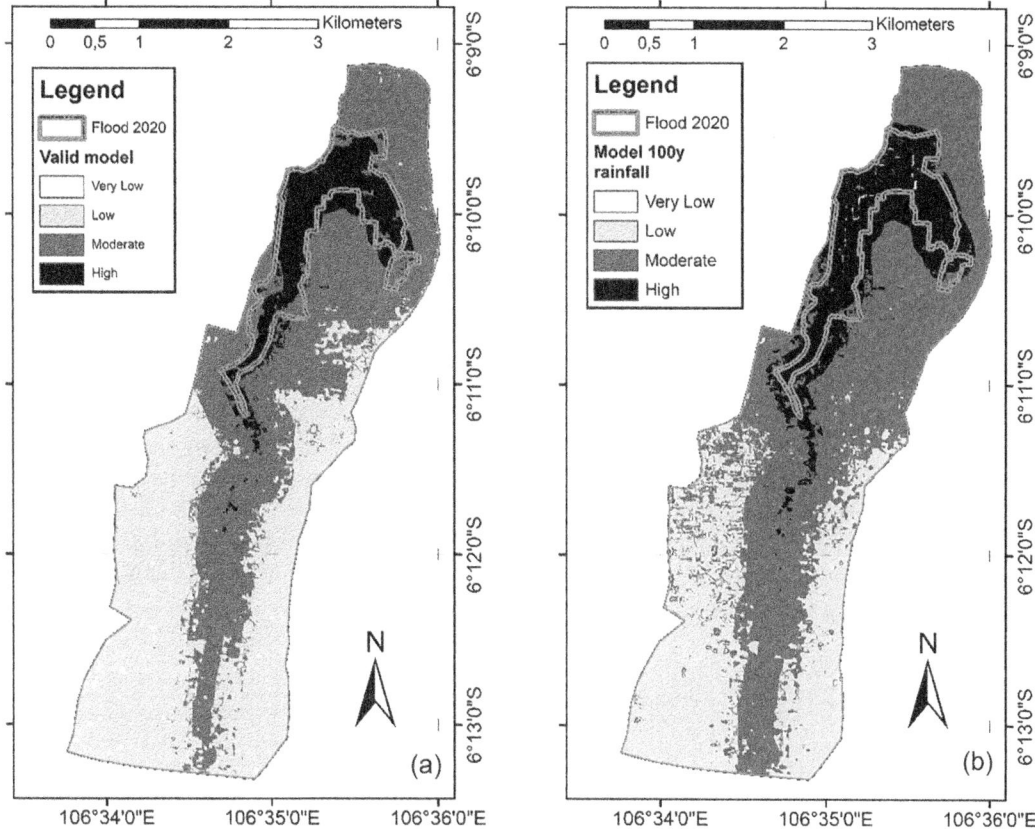

Fig. 7.2 Flood susceptibility map in the study area: (**a**) Valid model, (**b**) 100-year rainfall model. (Source: created by authors)

this period, the highly susceptible area increased from 1,233,334 m² to 1,639,065 m², representing a 32.9% increase. This information is significant as it can serve as a reference for local authorities in spatial planning decisions affecting land use patterns.

Precipitation data from several return periods were utilised to predict the rise in flood susceptibility in the study area. The flood susceptibility assessment generated a map for the *Kali* Ledug River basin, which illustrates varying levels of susceptibility across the region. The maps illustrate that regions with higher flood susceptibility are located in a broader floodplain, which has expanded in all directions compared to the inundation data from the flooding in February 2020. Visually, the area exhibiting high susceptibility to flooding is expanding due to the rise in precipitation. Based on this spatial analysis, the highly susceptible areas are predicted to increase as the precipitation intensity increases. Concerning precipitation levels expected to occur once every 100 years (Fig. 7.2b), the region exhibiting significant susceptibility would undergo an augmentation from 1,639,065 to 2,565,191 m², resulting in a rise of 57%. The significant increase in flood-affected areas must be a reference for local authorities in preparing facilities to deal with floods during emergency response.

7.3.2 Participatory Community Flood Resilience Analysis

Weighting using the AHP method resulted in satisfactory outcomes: the Flood Risk Perception Criterion and the Home Environment Criterion had the highest weight of 37.57% and 17.72%, respectively. Other criteria had varied weights: Communication and Information 11.06%, Institutional 9.99%, Social Capital 9.79%, Economic 7.09% and Social 6.79%. The Flood Resilience Indices were high for the three sub-districts, with 0.70, 0.77 and 0.74 for SD1, SD2 and SD3, respectively (Fig. 7.3). However, the indicator indices varied considerably among the sub-districts: SD1 had the highest social index; SD2 had the highest home environment, social capital and perception indices; and SD3 had the highest institutional index. SD1 and SD2 also had equal economic indices; SD1 and SD3 had similar social indices, classified as moderate, while SD2 had a low social index. Furthermore, the communication and information, social capital, institutional and risk perception indices were very high, except for the risk perception in SD1, which was high. The social, economic and home environment indices were moderate, except for the social index in SD2, which was low.

Therefore, these results suggest that improving the three indicators to high or very high levels could enhance the resilience index in the study area.

The in-depth interview with ten neighbourhood leaders provided insight into the conditions of the community affected by floods. This study describes how the community attempted to minimise the impact of floods in their area. The results showed that the community chose to stay in the area for several reasons, such as strategic location, low property prices, good social relations and hope for government action to address the flood problem. However, they also faced challenges living in an inadequate housing environment, such as basic housing infrastructure that did not consider future dynamics and land subsidence that disrupted drainage functions. To overcome these challenges, they utilised various sources of information on potential floods through door-to-door, mosque loudspeakers or

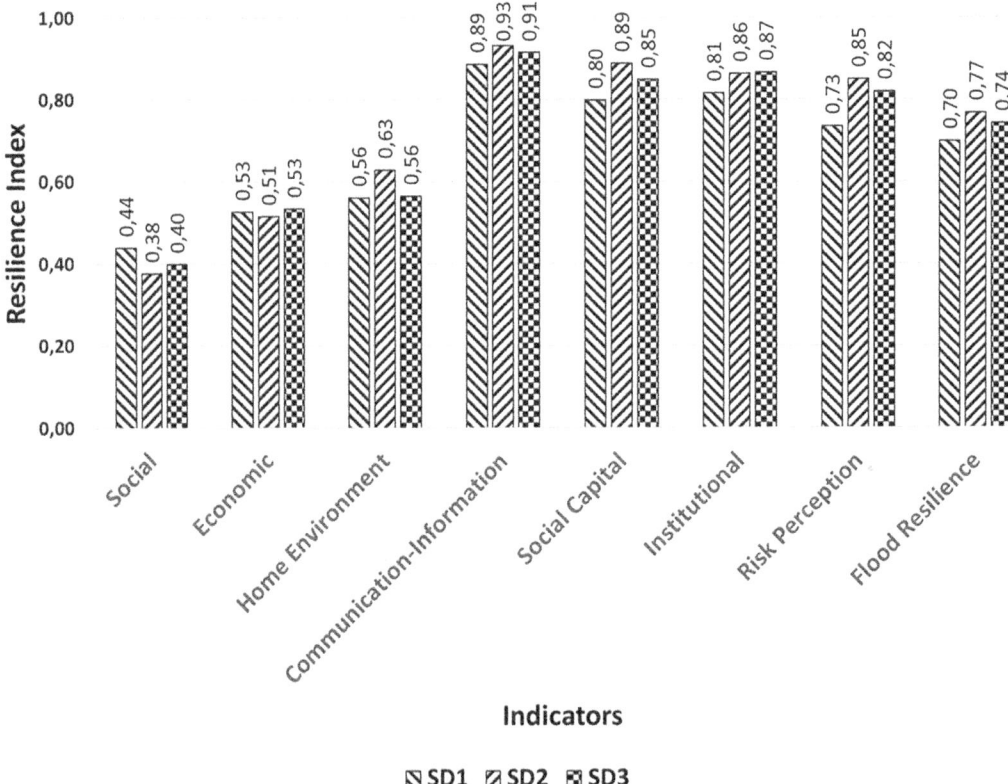

Fig. 7.3 Criteria and flood resilience index in each sub-district. (Source: created by authors)

social media based on official government notifications. This information helped them properly prepare for floods. They additionally relied on a strong social bond among community members as a foundation for mutual assistance among them. This social bond was forged through regular activities and informal interactions, despite differences in religion, ethnicity and socioeconomic status. United by a common purpose, they collaborated in responding to floods by establishing public kitchens, coordination centres and evacuation sites. Financial contributions to flood management were also made collectively. Moreover, they also received assistance from diverse stakeholders, including district, sub-district and local governments. The coordination among these parties was effective, facilitating the distribution of assistance through city authorities and local leaders. This study showed that the community could manage floods well in their area.

The people living in the study area were aware of the flood hazards due to alterations in the upper catchment and land subsidence. They also observed that the malfunctioning of drainage channels, gated weirs and pump stations caused by sedimentation and blockage could exacerbate flooding. Furthermore, they knew that building river embankments shifted flood risk from one area to another. Based on their understanding of the regularly occurring floods, the community developed adaptation strategies, such as raising buildings or adding floors, to reduce the impact of floods. For those who lacked funds, they created storage space in the ceiling to protect their belongings from floods. The community also cooperated to raise the road to be more accessible during floods. This study showed that the community had a high adaptation capacity to floods.

7.3.3 Transdisciplinary in Flood Resilience Study

This study ascertained that the land cover changes from 2001 to 2020 have increased the frequency and area of flood inundation by utilising a participatory approach to generate validation data. This study applied a participatory approach to analyse the flood susceptibility and determine the inundation areas in the river basin. The approach involved the flood-affected communities as validation sources for the developed geospatial models. By doing so, the study was able to incorporate their empirical experiences into the spatial analysis and enhance its accuracy and reliability. Furthermore, the participatory approach benefited the communities, as they received information on their area's flood susceptibility and inundation boundaries. This information increased their awareness of the flood risks and preparedness for potential floods in the rainy season.

This finding supports Samaddar et al. (2022), who found that participatory flood risk mapping requires community involvement to secure their livelihoods and empower marginalised groups in action plan formulation. A socio-cultural-based humanist approach in facilitation is also an important component that enhances community participation in Participatory Flood Risk Management (PERM) for Disaster Risk Reduction (DRR). This finding also supported Tiepolo et al. (2023), who found that local science contributed to the flood risk assessment regarding the critical daily rainfall threshold that caused pluvial flood damage. Data on flood-affected population records was used to verify the accuracy of the hydraulic model in simulating inundation areas. Tiepolo et al. (2023) produced outputs based on validation data generated using a participatory approach in their studies. Participatory research as part of a transdisciplinary approach using local knowledge of flood-affected communities should be implemented to enrich academia's analysis and modelling results. Local knowledge is held by many parties, including flood-affected households, neighbourhood leaders and local authorities. Incorporating the knowledge of all parties is very important to get a comprehensive picture of the flood risk in a specific location. However, applying a participatory approach could be a significant challenge. Samaddar et al. (2022) noted that the participatory approach required a lot of time and labour from local communities to coordinate the activities. Hence, social and cultural events that respect the diversity of regional traditions should be held to recognise the role of local leaders.

The results of geospatial modelling related to flood susceptibility can be used to increase public awareness of the flood risk they face and are expected to be a practical approach to raising awareness about specific flood hazards. This approach aims to equip communities in flood-prone regions with the necessary knowledge to prepare for potential flooding events. Awareness is one aspect that needs to be raised in the flood-affected community to make the actions taken to mitigate floods realistically. In many cases, due to short-term thinking considerations, solutions offered by the government, such as relocation programmes, are rejected by the flood-affected community for various reasons. This study expands upon the earlier study conducted by Puzyreva and de Vries (2021), which emphasised the importance of enhancing perceptions of flood risks and safety and highlighted the significance of employing effective communication strategies for flood risk mitigation.

Moreover, these research findings reinforce the conclusions drawn by Rollason et al. (2018), who emphasised providing information about potential floods to residents in areas highly susceptible to flooding. Flood susceptibility maps, which illustrate a range of rainfall scenarios spanning a 100-year recurrence interval, are invaluable tools for effectively communicating the potential flood hazards to the public. Additionally, the current study aligns with the recommendation proposed by Rollason et al. (2018) for flood risk managers to enhance community literacy by promoting local knowledge about the specific flood potentials of these communities. Promoting local wisdom and understanding of flood risks makes establishing a more resilient and well-prepared populace in these vulnerable areas possible.

Another finding is that a transdisciplinary approach can be used to confirm the data obtained from the questionnaire survey of the flood-affected communities. In this study, the opinions of the neighbourhood leaders were used as comparative data against the questionnaire survey results. This study applied a combination of methods to comprehensively analyse household flood resilience. The criteria of communication and information, social capital, institutional factors and flood risk perception had very high indexes, while social, economic and home environment had high scores. These factors formed a high-value flood resilience index in the study area. This study confirmed the applicability of previous research on subjective flood resilience assessment methods (Clare et al., 2017), which were still in the early stages of development. This study also aligned with Nguyen and James (2013), who examined flood resilience assessment using a subjective approach in the Mekong River Delta, Vietnam. Quantitative methods could be implemented quickly using questionnaire instruments. They could be easily replicated in other locations, while qualitative methods required more time and effort in the process of subjective reasoning in interpreting the findings and could not be easily replicated in other study areas. The transdisciplinary approach confirming the quantitative data from the questionnaire survey with the qualitative data obtained from the neighbourhood leaders gave a comprehensive understanding of flood resilience at the household and community levels.

This study demonstrated how a transdisciplinary approach in flood risk management can improve flood resilience by engaging stakeholders in different aspects of the problem. The following could be achieved with a balanced understanding of river basin flood susceptibility among all stakeholders. First, the local authorities and community could collaborate to improve the structural mitigation measures, such as dams, dikes or drainage systems, to reduce the potential occurrence of floods. Second, the community's social capital could be increased by strengthening the trust, solidarity and cooperation among the residents, which would form the basis for more intensive collaboration and mutual assistance when floods occur. This condition was predicted to increase community flood resilience. Third, the risk perception of the community could be increased by providing accurate and timely information about flood hazards and impacts. Fourth, communication and information sharing among the stakeholders could be increased using various channels and platforms such as social media, radio and newsletters. These third and fourth conditions would increase the community's preparedness in facing floods so that the level of loss could be minimised. Lastly, the institutional cooperation among multi-stakeholders could be improved by establishing clear roles, responsibilities and coordination mechanisms. If cross-stakeholder collaboration within this institutional framework is achievable, flood management measures could be carried out from the mitigation, preparedness, response and rehabilitation phases. The finding reinforces Hudson et al. (2020), who found that social capital and risk perception dominate flood resilience. In addition, this study supports S. Ali and George (2022) that communication is also essential in flood resilience. On the institutional aspect, this study is similar to Singh et al. (2021), who emphasised that the institutional factor is essential in shaping flood resilience. This study shows that a transdisciplinary approach in flood risk management can enhance flood resilience if communicated effectively to all stakeholders.

In this study, we have demonstrated that the authors, comprising persons from academia and practitioners, have applied a transdisciplinary approach using spatial methods to understand space's social and relational dimensions. This method is essential for understanding changes in flood susceptibility within the river basin area, which serves as a residence and a location for socio-economic activities. This study emphasised the significance of incorporating citizen knowledge as a critical component in analysing flood susceptibility and household flood resilience issues. Doing so gave us a more comprehensive understanding of the interaction between humans and their surrounding environment, providing a solid foundation for developing actionable flood risk management strategies. This spatial approach will serve as the basis for promoting urban sustainability and advancing the achievement of Sustainable Development Goal 11.

7.4 Conclusion

This study explored transdisciplinary flood risk management based on field practises. It assessed river basin flood susceptibility using a participatory geospatial approach with land cover data from 2001 to 2021 and examined community flood resilience in the study area, eliciting subjective viewpoints from flood-affected households and local leaders. The study has several findings: flood-affected communities participated in flood susceptibility analysis and inundation area determination, providing information for geospatial model validation and spatial analysis enrichment. The participatory approach also increased their awareness and preparedness for potential floods by providing information on their area's flood susceptibility and inundation boundaries. The study also revealed that a comprehensive understanding of the flood adaptation strategies of households and communities was achieved by combining the quantitative and qualitative data from the survey and the neighbourhood leaders in a transdisciplinary approach.

The last finding is that a transdisciplinary approach in flood risk management can improve flood resilience by engaging stakeholders in different aspects of the problem. This situation could achieve a balanced understanding of river basin flood susceptibility and lead to the following outcomes: improved structural mitigation measures, increased social capital, risk perception, communication and information sharing among the community, and enhanced institutional cooperation and coordination among the multi-stakeholders. These outcomes would reduce flood occurrence, impact and loss and increase the community's capacity to cope and recover.

A transdisciplinary approach is required to conduct a scientific assessment involving various academic disciplines, including geospatial analysis, remote sensing, water resource management, disaster risk management, social sciences, communication and public policy. Furthermore, incorporating local knowledge from flood-affected communities, neighbourhood leaders and local authorities will facilitate the development of implementable flood resilience strategies applicable at both the public policy and practical implementation levels. By involving communities and local authorities in implementing flood resilience strategies, the chances of success are higher or much better. The successful implementation of such public policy could ultimately be crucial in achieving Sustainable Development Goal 11, which emphasises the importance of reducing mortality, morbidity and economic loss, enhancing urban resilience to disasters like floods and promoting participatory approaches in urban planning and management.

References

Ali, S., & George, A. (2022). Modelling a community resilience index for urban flood-prone areas of Kerala, India (CRIF). *Natural Hazards, 113*(1), 261–286. https://doi.org/10.1007/s11069-022-05299-7

Almoradie, A., de Brito, M. M., Evers, M., Bossa, A., Lumor, M., Norman, C., Yacouba, Y., & Hounkpe, J. (2020). Current flood risk management practices in Ghana: Gaps and opportunities for improving resilience. *Journal of Flood Risk Management, 13*(4). https://doi.org/10.1111/jfr3.12664

Baptista, S. R. (2014). *Design and use of composite indices in assessment of climate change vulnerability and resilience* (Issue July). The Earth Institute, Columbia University.

Beccari, B. (2016). A comparative analysis of disaster risk, vulnerability and resilience composite indicators. *PLOS Currents Disasters, 8*, 1–57. https://doi.org/10.1371/currents.dis.19f9c194f3e3724d9ffa285b157c6ee3

Bertilsson, L., Wiklund, K., de Moura Tebaldi, I., Rezende, O. M., Veról, A. P., & Miguez, M. G. (2019). Urban flood resilience—A multi-criteria index to integrate flood resilience into urban planning. *Journal of Hydrology, 573*(June 2018), 970–982. https://doi.org/10.1016/j.jhydrol.2018.06.052

Bottazzi, P., Winkler, M. S., Boillat, S., Diagne, A., Sika, M. M. C., Kpangon, A., Faye, S., & Speranza, C. I. (2018). Measuring subjective flood resilience in Suburban Dakar: A before-after evaluation of the "Live with Water" project. *Sustainability (Switzerland), 10*(7), 2135. https://doi.org/10.3390/su10072135

Braun, V., & Clarke, V. (2006). Using thematic analysis in psychology. *Qualitative Research in Psychology, 3*(2), 77–101. https://doi.org/10.1191/1478088706qp063oa

Bui, D. T., Tsangaratos, P., Ngo, P. T. T., Pham, T. D., & Pham, B. T. (2019). Flash flood susceptibility modeling using an optimized fuzzy rule based feature selection technique and tree based ensemble methods. *Science of the Total Environment, 668*, 1038–1054. https://doi.org/10.1016/j.scitotenv.2019.02.422

Clare, A., Graber, R., Jones, L., & Conway, D. (2017). Subjective measures of climate resilience: What is the added value for policy and programming? *Global Environmental Change, 46*(October 2016), 17–22. https://doi.org/10.1016/j.gloenvcha.2017.07.001

Ekmekcioğlu, Ö., Koc, K., & Özger, M. (2021). District based flood risk assessment in Istanbul using fuzzy analytical hierarchy process. *Stochastic Environmental Research and Risk Assessment, 35*(3), 617–637. https://doi.org/10.1007/s00477-020-01924-8

Feloni, E., Mousadis, I., & Baltas, E. (2020). Flood vulnerability assessment using a GIS-based multi-criteria approach—The case of Attica region. *Journal of Flood Risk Management, 13*(S1), 1–15. https://doi.org/10.1111/jfr3.12563

Fratini, C. F., Geldof, G. D., Kluck, J., & Mikkelsen, P. S. (2012). Three Points Approach (3PA) for urban flood risk management: A tool to support climate change adaptation through transdisciplinarity and multifunctionality. *Urban Water Journal, 9*(5), 317–331. https://doi.org/10.1080/1573062X.2012.668913

Houston, D., Werritty, A., Ball, T., & Black, A. (2020). Environmental vulnerability and resilience: Social differentiation in short- and long-term flood impacts Donald. *Transactions of the Institute of British Geographers, 46*(1), 102–119. https://doi.org/10.1111/tran.12408

Hudson, P., Hagedoorn, L., & Bubeck, P. (2020). Potential linkages between social capital, flood risk perceptions, and self-efficacy. *International Journal of Disaster Risk Science, 11*, 251. https://doi.org/10.1007/s13753-020-00259-w

IPCC. (2022). Climate change 2022: Impacts, adaptation, and vulnerability. In H.-O. Pörtner, D. C. Roberts, M. Tignor, E. S. Poloczanska, K. Mintenbeck, A. Alegría, M. Craig, S. Langsdorf, S. Löschke, V. Möller, A. Okem, & B. Rama (Eds.), *Contribution of working group II to the sixth assessment report of the intergovernmental panel on climate change*. (3056 pp). Cambridge University Press. https://doi.org/10.1017/9781009325844

Jones, L., & Tanner, T. (2017). 'Subjective resilience': Using perceptions to quantify household resilience to climate extremes and disasters. *Regional Environmental Change, 17*(1), 229–243. https://doi.org/10.1007/s10113-016-0995-2

Li, Y., Gong, S., Zhang, Z., Liu, M., Sun, C., & Zhao, Y. (2021). Vulnerability evaluation of rainstorm disaster based on ESA conceptual framework: A case study of Liaoning province, China. *Sustainable Cities and Society, 64*(February 2020), 102540. https://doi.org/10.1016/j.scs.2020.102540

Marchezini, V. (2020). Transdisciplinary research as a support for the planning of disaster risk management actions. *Saúde Em Debate, 44*(spe2), 33–47. https://doi.org/10.1590/0103-11042020e203i

Mashi, S. A., Inkani, A. I., Obaro, O., & Asanarimam, A. S. (2020). Community perception, response and adaptation strategies towards flood risk in a traditional African city. In *Natural hazards* (Vol. 103(2), p. 1727). Springer. https://doi.org/10.1007/s11069-020-04052-2

Minucci, G., Molinari, D., Gemini, G., & Pezzoli, S. (2020). Enhancing flood risk maps by a participatory and collaborative design process. *International Journal of Disaster Risk Reduction, 50*(June), 101747. https://doi.org/10.1016/j.ijdrr.2020.101747

Mondal, M. S. H., Murayama, T., & Nishikizawa, S. (2021). Examining the determinants of flood risk mitigation measures at the household level in Bangladesh. *International Journal of Disaster Risk Reduction, 64*(July), 102492. https://doi.org/10.1016/j.ijdrr.2021.102492

Mukherjee, F., & Singh, D. (2020). Detecting flood prone areas in Harris County: A GIS based analysis. *GeoJournal, 85*(3), 647–663. https://doi.org/10.1007/s10708-019-09984-2

Narendra, B. H., Setiawan, O., Hasan, R. A., Siregar, C. A., Pratiwi, Sari, N., Sukmana, A., Dharmawan, I. W. S., & Nandini, R. (2023). Flood susceptibility mapping based on watershed geomorphometric characteristics and land use/land cover on a small island. *Global Journal of Environmental Science and Management, 10*(1), 1–20. https://doi.org/10.22034/gjesm.2024.01.07

Nguyen, K. V., & James, H. (2013). Measuring household resilience to floods: A case study in the Vietnamese Mekong River Delta. *Ecology and Society, 18*(3), 13. https://doi.org/10.5751/ES-05427-180313

Nsangou, D., Kpoumié, A., Mfonka, Z., Bateni, S. M., Ngouh, A. N., & Ndam Ngoupayou, J. R. (2022). The Mfoundi watershed at Yaoundé in the humid tropical zone of Cameroon: A case study of urban flood susceptibility mapping. *Earth Systems and Environment, 6*(1), 99–120. https://doi.org/10.1007/s41748-021-00276-9

Puzyreva, K., & de Vries, D. H. (2021). 'A low and watery place': A case study of flood history and sustainable community engagement in flood risk management in the County of Berkshire, England. *International Journal of Disaster Risk Reduction, 52*(November 2020), 101980. https://doi.org/10.1016/j.ijdrr.2020.101980

Räsänen, A., Kauppinen, V., Juhola, S., Setten, G., & Lein, H. (2020). Configurations of community in flood risk management. *Norsk Geografisk Tidsskrift, 74*(3), 165–180. https://doi.org/10.1080/00291951.2020.1754285

Rollason, E., Bracken, L. J., Hardy, R. J., & Large, A. R. G. (2018). Rethinking flood risk communication. *Natural Hazards, 92*(3), 1665–1686. https://doi.org/10.1007/s11069-018-3273-4

Saaty, T. L. (2008). Decision making with the analytic hierarchy process. *International Journal of Services Sciences, 1*(1), 83–98. https://doi.org/10.1504/IJSSCI.2008.017590

Sadiq, A. A., Tyler, J., & Noonan, D. S. (2019). A review of community flood risk management studies in the United States. *International Journal of Disaster Risk Reduction, 41*, 101327. https://doi.org/10.1016/j.ijdrr.2019.101327

Samaddar, S., Si, H., Jiang, X., Choi, J., & Tatano, H. (2022). How participatory is participatory flood risk mapping? Voices from the flood prone Dharavi Slum in Mumbai. *International Journal of Disaster Risk Science, 13*(2), 230–248. https://doi.org/10.1007/s13753-022-00406-5

Santosa, B. H., Martono, D. N., Purwana, R., & Koestoer, R. H. (2022). Flood vulnerability evaluation and prediction using multi-temporal data: A case in Tangerang, Indonesia. *International Journal on Advanced Science, Engineering and Information Technology, 12*(6), 2156–2164. https://doi.org/10.18517/ijaseit.12.6.16903

Santosa, B. H., Martono, D. N., Purwana, R., Koestoer, R. H., & Susanti, W. D. (2023). Understanding household flood resilience in Tangerang, Indonesia, using a composite indicator method. *Natural Hazards, 119*, 69. https://doi.org/10.1007/s11069-023-06120-9

Singh, P., Amekudzi-Kennedy, A., Woodall, B., & Joshi, S. (2021). Lessons from case studies of flood resilience: Institutions and built systems. *Transportation Research Interdisciplinary Perspectives, 9*(July 2020), 100297. https://doi.org/10.1016/j.trip.2021.100297

Snel, K. A. W., Priest, S. J., Hartmann, T., Witte, P. A., & Geertman, S. C. M. (2021). Do the resilient things 'Residents' perspectives on responsibilities for flood risk adaptation in England. *Journal of Flood Risk Management, 14*(3), 1–14. https://doi.org/10.1111/jfr3.12727

Sun, R., Shi, S., Reheman, Y., & Li, S. (2022). Measurement of urban flood resilience using a quantitative model based on the correlation of vulnerability and resilience. *International Journal of Disaster Risk Reduction, 82*(December 2021), 103344. https://doi.org/10.1016/j.ijdrr.2022.103344

Takara, K. (2018). Promotion of interdisciplinary and transdisciplinary collaboration in disaster risk reduction. *Journal of Disaster Research, 13*(7), 1193–1198. https://doi.org/10.20965/jdr.2018.p1193

Tate, E., Decker, V., & Just, C. (2021). Evaluating collaborative readiness for interdisciplinary flood research. *Risk Analysis, 41*(7), 1187–1194. https://doi.org/10.1111/risa.13249

Thaler, T., Clar, C., Junger, L., & Nordbeck, R. (2021). Opportunities and challenges for transdisciplinary research in flood risk management: Some critical reflections and lessons learnt for research on sustainability. *Journal on Protected Mountain Areas Research and Management, 13*(2), 42–47. https://doi.org/10.1553/0X003C9DA2

Tiepolo, M., Braccio, S., Fiorillo, E., Galligari, A., Katiellou, G. L., Massazza, G., & Tarchiani, V. (2023). Participatory risk assessment of pluvial floods in four towns of Niger. *International Journal of Disaster Risk Reduction, 84*(February 2022), 103454. https://doi.org/10.1016/j.ijdrr.2022.103454

Vardoulakis, S., Matthews, V., Bailie, R. S., Hu, W., Salvador-Carulla, L., Barratt, A. L., & Chu, C. (2022). Building resilience to Australian flood disasters in the face of climate change. *Medical Journal of Australia, 217*(7), 342–345. https://doi.org/10.5694/mja2.51595

Raldi Hendro Koestoer is a faculty member at the School of Environmental Science, Universitas Indonesia. He holds BA & Doktorandus in Geography from Universitas Indonesia, an MA in Regional Science from the University of Queensland and a PhD in Environmental Science from Griffith University, Australia. Having a Professor of LIPI, he has held various government positions, including Indonesian Senior Official for IMT GT and BIMP EAGA (2009–2018) and Principal Senior Policy Analyst (2018 to present). His research interests encompass urban and regional environments, spatial planning, sustainable development, spatial modelling and public policy.

Budi Heru Santosa specialises in water resource management, with a focus on flood risk management. He employs an interdisciplinary and transdisciplinary approach, integrating scientific analyses from various fields and non-academic knowledge. Additionally, he leverages advanced technologies such as Artificial Intelligence, Geographic Information Systems (GIS), and remote sensing to assess changes in river basin flood susceptibility resulting from land use and spatial planning policies. Currently, Budi Heru Santosa serves as Chair of the Indonesian National Committee for UNESCO Intergovernmental Hydrological Programme.

Open Access This chapter is licensed under the terms of the Creative Commons Attribution 4.0 International License (http://creativecommons.org/licenses/by/4.0/), which permits use, sharing, adaptation, distribution and reproduction in any medium or format, as long as you give appropriate credit to the original author(s) and the source, provide a link to the Creative Commons license and indicate if changes were made.

The images or other third party material in this chapter are included in the chapter's Creative Commons license, unless indicated otherwise in a credit line to the material. If material is not included in the chapter's Creative Commons license and your intended use is not permitted by statutory regulation or exceeds the permitted use, you will need to obtain permission directly from the copyright holder.

Bridging the Gap Between Academia, Practitioners and Communities: A Transdisciplinary Process Towards Regenerative Public Space in South Africa

Karina Landman and Ilan Guest

8.1 Introduction

As the world is rapidly changing, cities are confronted with new challenges to deal with these changes in ways that will consider the future well-being of the planet and its people. To do so, public spaces need to be able to adapt and transform to address challenges of rapid urbanisation, densification, climate change, social conflict, exclusion and disconnection from nature. The Sustainable Development Goals (SDGs) offer some direction to do so. For example, SDG11 sets out to 'make cities and human settlements inclusive, safe, resilient and sustainable', while Goal 11.7 specifically targets the transformation of public space. The goal is to 'by 2030, provide universal access to safe, inclusive and accessible, green and public spaces, in particular for women and children, older persons and persons with disabilities.' Many scholars have offered different processes and examples of how to achieve this through research and practise. This chapter, while supporting a holistic approach to sustainability, offers an example of transdisciplinary research for spatial planning and design related to SDG11.7.

The thinking about sustainability has also evolved over the years, alongside debates about research methods for urban sustainability. While many organisations such as the UN and the World Bank promote the development of sustainable settlements and public spaces, the thinking about sustainability also progressed from conventional and contemporary sustainability, including resilience thinking, to regenerative sustainability (Reed, 2007; Du Plessis, 2012, 2022; Mang & Reed, 2012; Benne & Mang, 2015; Hes & Du Plessis, 2014; Gibbons, 2020). In addition, many authors are calling for transdisciplinary research to better understand and investigate sustainability (Lang et al., 2012; Schauppenlehner-Kloyber & Penker, 2014; Polk, 2015; Balsiger, 2015; Leemans, 2016). At the same time, authors expressed the need for sustainability research to focus on a greater connection with practise and with communities in different types of settlements (Franklin & Blyton, 2011; Marsden, 2011; Mang et al., 2016). Transdisciplinary research is promoted as a mode of knowledge production that is effective to address sustainability challenges (Polk, 2015) and to deal with complexity in architecture and urbanism (Doucet & Janssens,

K. Landman (✉)
Department of Town and Regional Planning, University of Pretoria, Pretoria, South Africa
e-mail: karina.landman@up.ac.za

I. Guest
SATPLAN ALPHA, Johannesburg, South Africa
e-mail: ilan@satplan.co.za

2011). Given this, it raises questions about the spatial methods that would be relevant to investigate urban sustainability through transdisciplinarity in cities and smaller parts, such as public spaces. Spatial methods, in the context of this chapter, are considered as research methods that are applied to spatial phenomena and questions to enable spatial planning and design in practise.

This chapter concentrates on the City of Tshwane and highlights how different stakeholders work together towards regenerative public space in four different areas in the broader metropolitan area. The discussion shows the connection between regenerative sustainability (Hes & Du Plessis, 2014), transdisciplinary research spatially orientated methods and then foregrounds a project on regenerative public space to demonstrate how a group of stakeholders from academia, the private sector and communities, including architects, planners and GIS specialists, are working together across disciplines and beyond academia towards the transformation of public space.

8.2 The Evolution of Sustainability Thinking and Its Implications for Transdisciplinary Spatially Orientated Research

The thinking about sustainability evolved over the years. Sustainable development as a concept emerged from a growing awareness of the relationship between environmental problems and socio-economic challenges, such as poverty and inequality, raising serious questions about the concerns of a healthy future for humanity (Hopwood & Mellor, 2005). The movement grew due to the realisation that the actions of humans have an impact on the environment and that the resources we rely on are limited (Du Plessis, 2022). The concept originated to describe conserving environmental resources for human benefit or to mitigate the depletion and degradation of resources (Benne & Mang, 2015) and balance humans' needs with the ecological carrying capacity. The Bruntland report defined sustainable development as 'meeting the needs of the present without compromising the ability of future generations to meet their needs' (WCED, 1987: 43). This defines needs from a human standpoint and sustainable development as an anthropocentric concept (Hopwood & Mellor, 2005). Therefore, despite the realisation that humans are dependent on nature, the focus is anthropocentric and based mainly on how to continue economic growth within the context of finite resources (Gibbons, 2020). Given this, traditional sustainability focused on greening, conservation, the minimisation of scarce resources and overall, an attempt to mitigate human-induced changes and minimise the damage. In other words, the focus of traditional sustainability has been on more effective performance and as noted by Du Plessis (2022), improved functionality in cities and with relation to the built environment.

The need to work with change and consider the broader system paved the way for a revised sustainability paradigm. Contemporary sustainability moves beyond a focus on economic growth through notions of ecosystem viability, social justice, socio-ecological systems and satisfying livelihoods. Although the focus is still anthropocentric, thinking moved beyond only economic gains to human well-being and the acknowledgement that complex problems are value-laden and locally specific (Gibbons, 2020). These debates highlighted the importance of systems thinking and the need to consider the interconnected nature of socio-ecological systems (Capra, 2016) and the city as a socio-ecological system (Du Plessis, 2012). The focus on complex socio-ecological systems also introduced notions of transition, risk, vulnerability and resilience (Gibbons, 2020) and hence focus on resilience thinking. A resilience-based approach attempts to learn how to respond and adapt to change while avoiding situations that would move local and global socio-ecological systems closer to tipping points threatening their life-supporting and life-enhancing capacity (Du Plessis, 2012:17). The importance of relationships is critical to understand the various inter-linked systems operating in different types and parts of settlements and how success or failure in one system or part can have consequences for the other or broader system, especially in rapidly changing cities. Interventions to improve the spatial resilience of the urban environment tend to mainly focus on the urban form (Du Plessis, 2022). While contemporary sustainability recognises the value of systems thinking and offers various mechanisms to conceptualise and work with rapid change in cities, the focus often remains on adaptation and the restoration of broken systems.

Regenerative development and design extend the work on sustainability and resilience. Its proponents argue for a paradigm change from a 'mechanistic' to an 'ecological' or living systems worldview. An ecological paradigm considers the entire social-ecological system (SES) across many scales. Such an approach requires an understanding of multiple dimensions—the technical and environmental, as well as the psychological and spiritual to reconsider humans as an integral part of all living systems, which are part of the complex web of life across various scales, ranging from the global to the very local (Du Plessis, 2012; Benne & Mang, 2015; Mang et al., 2016). A focus on interdependent living systems requires a move from anthropocentric to autopoietic, creating a system capable of producing and supporting mutual benefit and symbiosis between all parts of the living system (Du Plessis, 2012; Gibbons, 2020). This implies an evolution from humans doing things to nature (restorative) to humans being an integral part of nature (reconciliatory) and, finally, humans participating as nature through co-evolution of the whole system (regenerative) (Reed, 2007: 676). It also shifts the focus from the function and form of the built environment to emphasise the flows (Du Plessis, 2022). In this regard, urban planning and design would enable the various flows of people, goods and information through different types of connections and points of exchange to create life-supporting networks for living systems to flourish.

Understanding the evolution of sustainability is important to identify different approaches to sustainable development by various organisations and stakeholders. However, one needs to keep in mind that these various ways of thinking are not mutually inclusive but that regenerative sustainability builds on and incorporates the previous notions whilst taking the debates forward.

As mentioned, there has been a call for transdisciplinary research to deal with the complexities related to researching sustainability in cities or parts thereof (Doucet & Janssens, 2011; Polk, 2015). Cross-disciplinary research refers to types of approaches to research that involve more than one discipline at a time. This can occur in different ways; hence the distinction between multidisciplinary, interdisciplinary and transdisciplinary research. These approaches are complementary rather than being mutually exclusive, because without specialised disciplinary studies there would be no in-depth knowledge and data (Lawrence, 2004).

Multidisciplinary research refers to a case/project where each researcher remains within their discipline and contributes using disciplinary concepts and methods (Lawrence, 2004) and works in a self-contained manner with little or no cross-fertilisation or synergy in the outcomes (Bruce et al., 2004). Interdisciplinary contributions bring together different disciplines, which retain their own concepts and methods applied to a mutually agreed subject or project (Lawrence, 2004), while researchers work in a collaborative way to ensure a common understanding (Balsiger, 2004). The contributions from diverse disciplines are integrated to provide a holistic or systemic outcome (Bruce et al., 2004).

Transdisciplinary approaches are more comprehensive in scope and vision (Klein, 1990) and imply a fusion of disciplinary knowledge that creates a new hybrid that is different from any specific parts—requiring 'transcendence' (Lawrence, 2004). Transdisciplinarity involves conceptual frameworks that transcend the narrow scope of disciplinary world views, 'breaking through transdisciplinary boundaries' (Klein, 1990:66) to organise knowledge around complex heterogeneous domains (Bruce et al., 2004). Collectively, transdisciplinary contributions enable the cross-fertilisation of ideas and knowledge from various contributors that leads to an enlarged vision of the subject, as well as new explanatory theories. Transdisciplinarily is a way of achieving innovative goals, enriched understanding and a synergy of new methods, requiring a common conceptual framework and analytical methods based on shared terminology, mental images and common goals (Lawrence, 2004). In this context, the outcome of transdisciplinary research has to do with 'knowledge coherence' as opposed to 'knowledge unity', which is the result of interdisciplinary research (Ramadier, 2004). Following this, transdisciplinarity can be defined as '…a reflexive, integrative, method-driven scientific principle aiming at the solution or transition of societal problems and concurrently of related scientific problems by differentiating and integrating knowledge from various scientific and societal bodies of knowledge' (Lang et al., 2012:26–27).

Urban sustainability is highly complex and contested, involving conflicting political agendas. By creating hybrid spaces and communities of co-production through transdisciplinary research, it is possible to develop new ways of working together. Polk (2015) furthermore highlights five focal areas to assist with implementing transdisciplinary research and address urban challenges related to sustainability. These are inclusion, collaboration, integration, usability and reflexivity. Inclusion calls for groups from both practise and research, while collaboration implies that both these groups should offer in-depth contributions. Integration refers to a synthesis of both practise-based and scientific perspectives, values and knowledge, while usability implies the social robustness and transformative capacity of the outputs and outcomes. Finally, reflexivity calls for on-going scrutiny of the choices that are made when identifying and integrating diverse values, priorities and knowledge from both practise and science in the research process (Polk, 2015: 144). Given this, transdisciplinarity offers a way to engage with regenerative sustainability research and practises in a holistic way that would involve a range of disciplines across multiple dimensions and involve a wide range of stakeholders from academia, practise and communities.

Finally, with spatial intervention as the goal, transdisciplinary research for sustainability may incorporate a range of research methods that could be applied to spatial phenomena and questions. In line with regenerative thinking, the purpose of these methods would be to understand the various flows of people, goods and information through different types of connections and points of exchange to create life-supporting networks for living systems to flourish. A number of methods could be suitable to facilitate the inclusion, collaboration, integration, usability and reflexivity, including stakeholder workshops, focus groups focusing on spatial issues, action research, site observation and spatial analysis and mapping through GIS. The next section outlines how a project on regenerative public space in South Africa utilised a transdisciplinary approach and included members from academia, practise and communities.

8.3 Regenerative Public Space in the City of Tshwane

The City of Tshwane is a large municipal area located in the Gauteng Province in north of South Africa. The municipal area includes Pretoria and covers an extensive area of 6345 km^2. In 2017, the population was 3.31 million, as the municipal area housed 6% of the national and 25% of the provincial population. The principal economic activities included government and community services (30%), finance (25%) and manufacturing (13%). The municipality hosts over 135 foreign missions and organisations, has the highest concentration of medical institutions per square kilometre in South Africa and is considered as the knowledge and research and development capital of South Africa. In addition, in 2017 Tshwane contributed to 28.4% of Gauteng's GDP and 10% of national GDP, while 53% of transport equipment exported from South Africa originates in this municipal area. However, despite the concentration of economic and research activities, unemployment was estimated to be around 24% in 2017. In addition, there are a growing number of informal settlements with an estimated 19% of the population in 2017, residing in informal dwellings (City of Tshwane 2020/2021 IDP). The economic conditions have since deteriorated due to the Covid-19 pandemic, contributing to rising levels of inequality, poverty and unemployment and challenges related to service delivery and urban management and maintenance.

In 2022, a team from the University of Pretoria and planners from a private planning firm, SATPLAN ALPHA, embarked on a 3-year project on Regenerative Public Space (RPS) and focused on the City of Tshwane as the case study area. The RPS Project is funded by the National Research Foundation (NRF) in South Africa. The intention of the project is to recognise the true value of public space in cities, especially in South Africa, and contribute to uplifting changes in these spaces. Public space is considered to be spaces owned by the state and accessible to the public. To do so, public spaces need to be able to adapt and transform to address challenges of rapid urbanisation, densification, climate change, social conflict, exclusion and disconnection to nature. Through transdisciplinary cooperation and various interactive methods aimed at understanding and working with spatial transformation, the project intends to take forward the Sustainable Development Goals (SDGs) through an exploration of the regenerative potential of public space. The aim of the project is to articulate a process towards regenerative public space and develop a digital platform to facilitate its implementation. The project consists of three key phases taking place over three years. Phase 1 has been completed, Phase 2 partially and Phase 3 will be carried out in 2024.

8.3.1 Phase 1: Common Understanding and Framework (2022)

Phase 1 of the project focused on the development of a common conceptual framework and methods. This involved a number of project meetings, which included all the members of the team. The team consists of architects, urban planners, GIS/IT specialists with a built environment background and a theologian who works with marginalised urban communities, such as homeless people. The team used the first project meetings to debate the nature of regenerative public space to start orientating the different team members and find a common ground. This was important to develop a shared terminology, mental images, common goals and refine the research methods to be used. In line with the ethos of regenerative development and design, Phase 1 had to involve creating common ground and understating the story of place. Consequently, the team selected four areas (Figs. 8.1 and 8.2) in the municipal area to work with during the project. These areas were selected based on their ability to show various phases or notions of regeneration in practise.

The project utilised several research methods applied to understand the socio-spatial conditions in the different types of spaces based on a common understanding of regenerative public space. Regenerative space refers to spaces that would be able to continuously evolve to allow humans and nature to heal and thrive in those spaces. The project meetings were followed by a formal workshop, where the team invited several relevant stakeholders from academia, the public sector (municipalities), private practice and representatives of the four communities. The aim of the workshop was to create a common understanding of regenerative public space and the process. The workshop featured two presentations by a leading academic working on regenerative development and design and a practising architect/urban designer who completed a course on regenerative design and is trying to find ways to incorporate this in practise. The afternoon session included four break-away sessions facilitated by team-members that included a mix of stakeholders to discuss various issues related to the definitions and process towards regenerative public space, possible challenges and opportunities and the value of a digital tool to assist with the implementation thereof in practise. The afternoon was concluded by feedback from the various groups and a panel discussion.

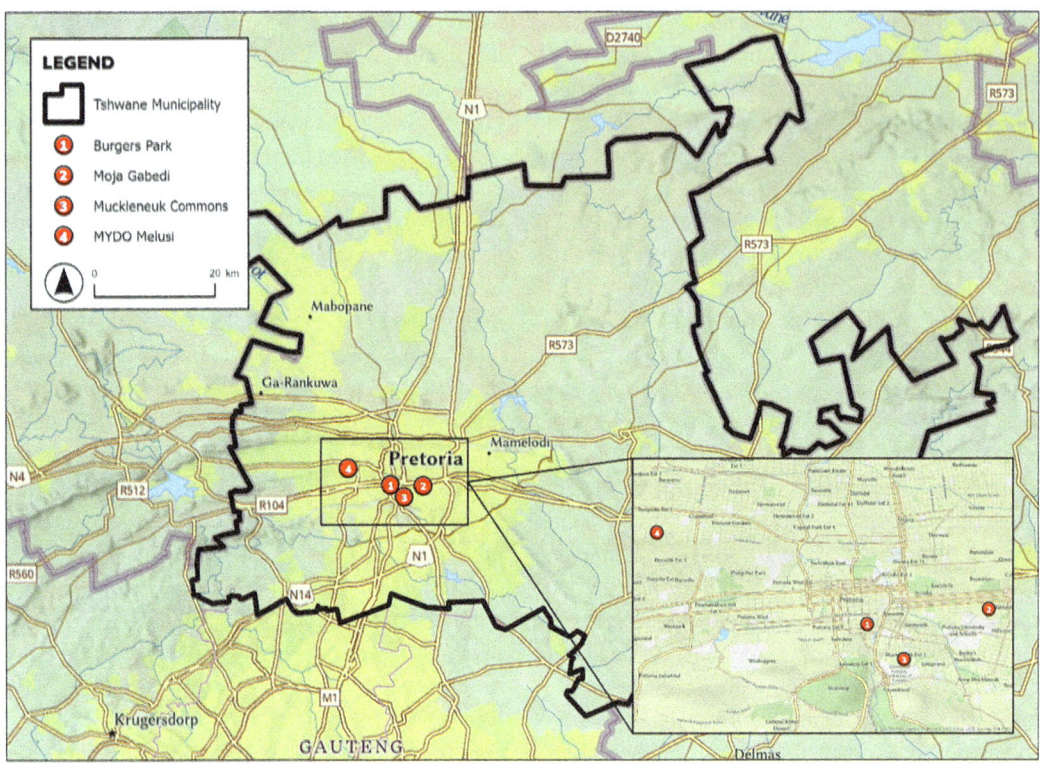

Fig. 8.1 Location of the study areas in the City of Tshwane. (Source: authors)

8.3.2 Phase 2: Understanding the Public Spaces and the Digital Platform

Phase 2 of the project focused on the development of case study content for the RPS process guides and the digital platform. This was initiated by four community workshops in each of the selected areas. Members of the team met with the key stakeholders on-site in each of the neighbourhoods with the intention to find out what happened in the spaces, why and what they are planning for the future. The various groups were asked to talk about the story of place and the process of finding the potential of the specific public space. They were also asked to elaborate on the process of transformation—identifying skills of and involving key people, developing action plans and gradually finding ways to implement the visions for the various spaces. Finally, they were asked to reflect on where they were now and where they wanted to go in the future. The site visits also included walking around with the community members, discussing existing or envisioned interventions and collecting spatial data through photographs.

- *Muckleneuk Commons*

The first community workshop took place in the Muckleneuk Commons in January 2023. Muckleneuk is an older, upmarket suburban neighbourhood close to the inner city of Pretoria. The community members shared their passion for nature but pointed out that the space was neglected and unsafe in the past. To improve and connect the community, they formed a neighbourhood association and developed an urban design framework to identify areas to promote safe and community-friendly public spaces. However, they faced a backlash from some community members who were concerned about attracting drug-users, loud music and alcohol abuse in the public space. The representatives and residents were responsible for drafting the

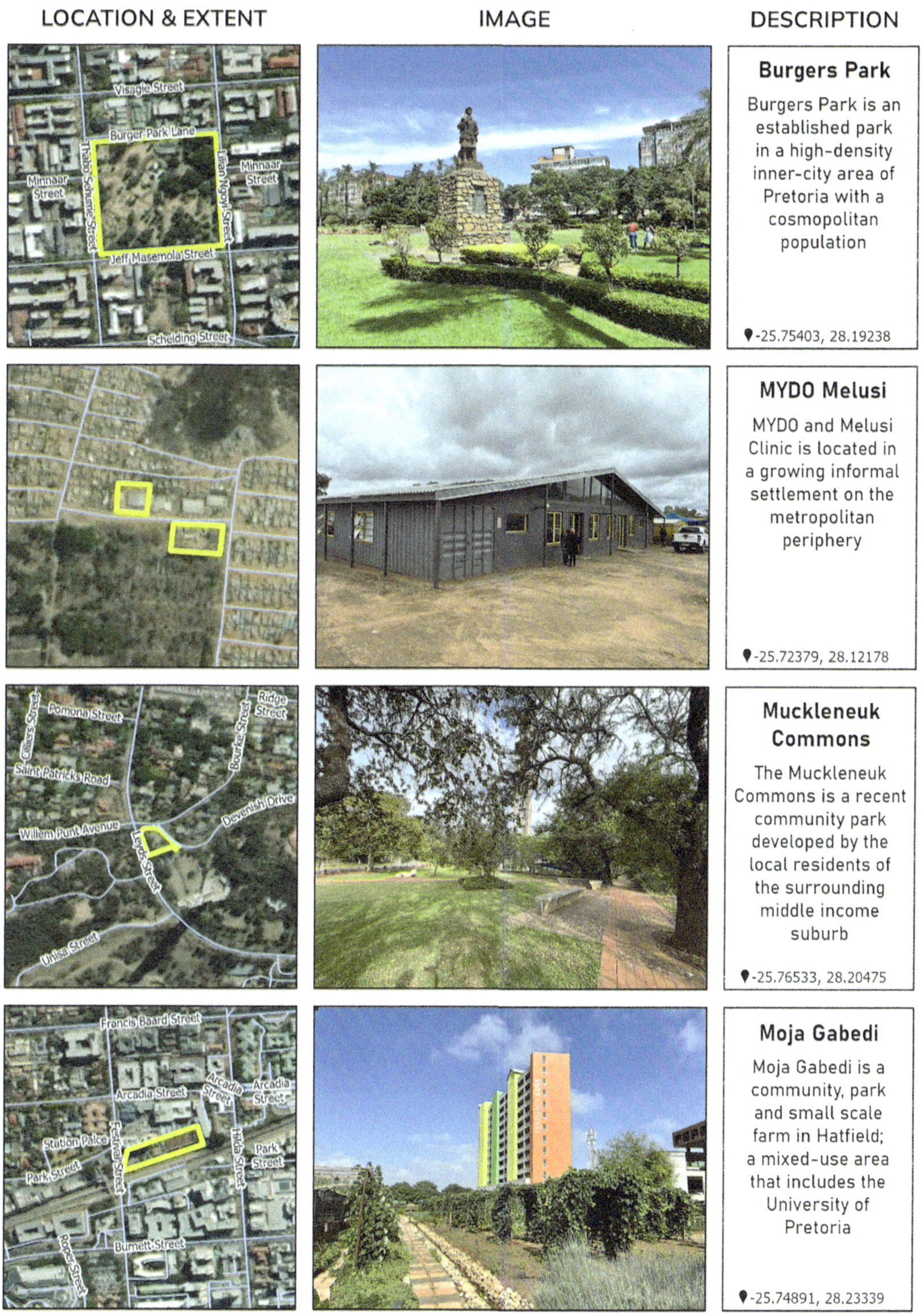
LOCATION & EXTENT
IMAGE
DESCRIPTION
Visagie Street
Burger Park Lane
Thabo Sehume Street
Lilian Ngoyi Street
Minnaar Street
Minnaar Street
Jeff Masemola Street
Scheiding Street
Burgers Park
Burgers Park is an established park in a high-density inner-city area of Pretoria with a cosmopolitan population
-25.75403, 28.19238
MYDO Melusi
MYDO and Melusi Clinic is located in a growing informal settlement on the metropolitan periphery
-25.72379, 28.12178
Cilliers Street
Pomona Street
Ridge Street
Bourke Street
Saint Patricks Road
Devenish Drive
Willem Punt Avenue
Leyds Street
Unisa Street
Muckleneuk Commons
The Muckleneuk Commons is a recent community park developed by the local residents of the surrounding middle income suburb
-25.76533, 28.20475
Francis Baard Street
Arcadia Street
Arcadia Street
Arcadia Street
Station Palce
Festival Street
Hilda Street
Park Street
Park Street
Burnett Street
Roper Street
Moja Gabedi
Moja Gabedi is a community, park and small scale farm in Hatfield; a mixed-use area that includes the University of Pretoria
-25.74891, 28.23339

Fig. 8.2 The four case study areas selected for the RPS Project. (Source: Authors) design, and physical elements were added in different phases as needed. Resources were obtained through donations, connections and community contributions. Telkom (a large telecommunications organisation), located near the commons, donated a significant amount and provided electricity and water. Maintenance and combating challenges like vandalism and misuse required constant involvement and active maintenance efforts. The community funded the maintenance through social activities and fundraisers, although financial challenges arose due to the impact of Covid-19. The representatives emphasised the importance of giving the public space value through ongoing community activity.

- *Melusi Open Spaces*

The second community workshop was held early in February 2023 in the informal settlement of Melusi. This settlement has a very high density with approximately 30,000 residents and 900 households. The area includes inadequate service provision and many of the housing units are informal. The workshop began with the community representatives from MYDO providing background on their settlement and its origins. They highlighted the community's growth due to the need for schools, as many students lived with their migrant working parents. MYDO was created to assist the youth with their schoolwork, providing food and extracurricular activities focused on sports and art programmes. They explained how they coordinated with community health workers and stakeholders to provide resources and health services to the youth. The representatives discussed the challenges faced by the community regarding service delivery and maintenance of services. They highlighted how MYDO and its centre became a prominent public space and community centre, with infrastructure sponsored by stakeholders and NGOs connecting them with more supporters. The MYDO Project has become the heart of the community, and the organisation is registering as an NGO to facilitate involvement from sponsors and stakeholders. They assist around 200–300 students and aftercare learners per day. Opposite to the clinic is a municipal nursery that was abandoned by the municipality and could be regenerated as a green open space and sports grounds. MYDO has expressed interest to initiate such a process.

- *Moja Gabedi*

The third workshop took place later in February 2023. Moja Gabedi is located on an open space situated in a dense and rapidly growing area of Tshwane, namely, Hatfield. The neighbourhood is one of the most dominant nodes in the City of Tshwane and is not only home to the University of Pretoria but also includes prominent schools, large corporations and office blocks, and higher-density, mixed-use housing developments. The representatives from Moja Gabedi and Reliable House, an institution focusing on homelessness, recognised the urgent need for addressing the negative spaces in the community. The strong influence of churches and religions in the area played a significant role in mobilising the community. Churches organised prayer sessions, providing food to participating community members. Reliable House offers homeless individuals access to food, prayer and accommodation, while Moja Gabedi, a garden in the city, provides working opportunities for recovering drug addicts from Reliable House. In addition, it is an open space that any member of the public can use. Over a period of 5 years, the space was transformed from a neglected and vacant open space to a vibrant, luscious garden. The representatives mentioned that resources, such as plants and physical structures, were donated by community stakeholders and integrated by the Moja Gabedi staff and volunteers from the community and the University of Pretoria. Unfortunately, the municipality had limited interest and involvement in the project, posing challenges related to land ownership for Moja Gabedi and Reliable House.

- *Burgers Park*

The final workshop was concerned with Burgers Park in the Inner City and took place in March 2023. The park is one of the oldest parks in the inner city of Pretoria, serving a very diverse population in a high-density and mixed-use area. The Victoria Garden of 47,678 m^2 was completed in 1874. One of the key stakeholder groups involved is *Friends of the Park*, which focuses on mobilising resources, engaging stakeholders and ensuring security in and around the park through collaboration with local and private law enforcement entities. They work with entities and institutions surrounding the park, including schools, churches and businesses, as well as those affected by it. Additionally, they try to foster a sense of ownership within the community and advocate the community's goals, needs and ideas to the city of Tshwane. The primary organisation overseeing several social programmes is the Tshwane Leadership Foundation (TLF), which manages and repurposes buildings in the area to address social needs, including housing for women, young girls, homeless individuals and those with chronic illnesses. They are assisted by Urban Studio, a partnership between the representatives, the Centre for Faith and Community, and the University of Pretoria. The Urban Studio focuses on providing social housing for homeless community members and women escaping from Gender-Based Violence situations. The participants highlighted the significance as a heritage park that offers access and opportunities for everyone. In the past, the park thrived as a tourist destination with restaurants and kiosks, particularly during the 90 s and the 2010 World Cup. However, since 2015 the park has faced decline due to reduced security and maintenance, resulting in the closure or abandonment of amenities such as the pond, nursery and public bathrooms. Efforts by the TLF to restore these facilities have been limited by the City of Tshwane. The representatives identified lack of ownership and safety as the major challenges faced by Burgers Park. The representatives acknowledged that their interaction with the park and the scheduling of events has decayed significantly, and they highlighted the lack of lighting at night as a safety concern. When discussing the park's decline, the representatives attributed it to reduced maintenance from the city, lack of ownership from the community and stakeholders, and the social issues of safety, drug misuse and homelessness. These groups aim to regenerate the park. The four community workshops not only highlighted the history and story of the four spaces but also the spatial nature of these spaces and their surrounding environments.

- *Digital Platform*

Following the individual community workshops, the project team discussed the nature and purpose of the digital platform with a view towards identifying various support areas and resources that may benefit their respective causes. The aim of the digital tool is to (1) share information related to the regenerative process and the four focus areas; (2) offer tools for communities to start their own processes in different public spaces; and (3) share information between built environment professionals and communities on an ongoing basis. These user group focus sessions were complemented by an international review of GIS-enabled digital platforms in the domains of public space, renewal and collaboration to better inform and focus the functionality of the RPS digital platform.

In recent years, GIS has evolved from primarily being a desktop application software into an Internet-based modular system of geospatial applications that are compatible with a wider set of devices and operating systems. Such Internet-based systems are configured as 'digital platforms' that can be customised for specific purposes. GIS digital platforms are also sometimes referred to as 'Online GIS' or 'Web GIS', whereby GIS tools and applications are hosted and used in an online environment. GIS digital platforms make use of 'the cloud' (i.e. servers that are accessed over the Internet, and the software and databases that run on those servers) to enable user access to GIS web apps, tools and mapping through an Internet browser. GIS digital platforms are powered by web services—standard services that deliver data and capabilities and connect components. The 'platform' aspect, sometimes referred to as the 'portal', provides a framework for sharing and using maps, apps and data. It also supports identity and provides the required infrastructure to manage users, and how they collaborate.

The modular components of GIS functionality (i.e. 'tools') that are configured into a GIS digital platform are referred to as 'GIS Web Apps'. There are a wide variety of GIS Web Apps that are designed for different applications, e.g. (1) Web-based mapping/GIS viewers—online interactive GIS maps; (2) Content repositories—spatial data libraries; (3) GIS editors—spatial data editing tools; (4) Interactive dashboards—dynamic dashboards of spatial data; (5) Online story maps—content/narrative driven map packages; and (6) Mobile GIS applications—field surveying tools. GIS digital platforms are custom purpose-built combinations of GIS Web Apps that are drawn together on a common platform and accessed via the Internet.

The digital platform of the RPS Project serves to provide public access to project-related content and resources. The platform is hosted online and seeks to leverage on a variety of geospatial tools and functionality that are available through the ESRI ArcGIS Online (AGOL) software suite.

The RPS digital platform has four components:

1. A digital library containing regenerative public space process guides, media and resources.
2. A story map of the RPS Project case studies.
3. An interactive webmap of RPS projects in the Tshwane area and several key base datasets.
4. A location capture tool and classification system for new RPS projects to be added to the RPS Projects register.

The spatial functionality of the RPS platform has been designed to promote the following outcomes:

1. Create exposure for RPS projects by making their location and information visible on a public access interactive webmap.
2. Provide public access to key base datasets such as cadastral information, registered public spaces and local ward council boundaries.
3. Create a system whereby new regenerative public space initiatives can be added to the RPS Projects register by the public via the interactive webmap and classified according to their particular themes.

- By focusing on RPS Project exposure and providing key base datasets alongside regenerative public space resources, the ultimate goal of this first version of the RPS digital platform is to create a 'common' where various RPS stakeholders, specialists and officials can find one another and interact in a transdisciplinary manner.

8.3.3 Phase 3: Testing and Launching the Digital Platform

Phase 3 of the project (2024) will include two additional workshops. The intention of the first workshop will be to obtain the input of the four communities on the digital platform and various tools. The second workshop will invite a wide range of stakeholders from other academic institutions, various government sectors, the private sector and various communities to discuss the value, use and ongoing development of the digital tool to assist with the implementation of regenerative public space in practise.

8.4 Blurring the Boundaries to Facilitate Deeper Levels of Collaboration, Participation and Cooperation

The discussion demonstrated how various research methods were applied to understand the spatial phenomenon of public space through a transdisciplinary approach to facilitate regenerative sustainability. This occurred in terms of three factors. First, the project enabled members from various disciplines to work together to develop a common conceptual framework and methods, fused together from various disciplines and emerging from ongoing discussions at project meetings and interactive multi-stakeholder workshops. This process allowed for deeper integration between the various disciplines and thus a form of reflexive transdisciplinarity achieved through the achievement of cognitive synthesis (Balsiger, 2015).

Second, the project involved members from academic, private practice and communities with various backgrounds. This allowed for citizen engagement in creating visions for urban planning and citizen involvement in the co-production of solutions for urban planning problems (Polk, 2015). The process also allowed for a broader collaboration and integration, designating a move towards hard transdisciplinarity or what has been described as the ideal-typical transdisciplinary process described by Lang et al. (2012). The community workshops not only allowed the project team to become familiar with the unique story of each neighbourhood and community but also to start identifying common patterns related to the different processes and spatial trends. In addition, it allowed for the inclusion of all the stakeholders, from academia, private practice and the communities to collaborate, as well as integrate both practise-based and scientific perspectives, values and knowledge, which are key focal areas defined by Polk (2015) to create hybrid spaces and allow for co-production of knowledge. The production of knowledge is enriched through a variety of actors, whether NGOs or residents from various neighbourhoods, so that the complexities are exposed by bridging the gap between theory and practise (Cupers et al., 2022). As the authors explain:

> Without theory, practise is unaware of its—often uncomfortable—blind spots and repercussions, while without practise, theory remains too easily dislocated, removed from its responsibilities toward the public. In building in and between theory and practise, we provide various trajectories, bridges across which we connect different spatial spheres, modes of knowledge production and disciplinary approaches, infusing theory with practise, and vice versa. (Ibid, p. 153)

Incorporating a mix of approaches and methods, such as engaging various city partnerships and city spaces with different actors and in a wide range of contexts, shows the various ways in which theory and practises are entangled (Ibid, p. 155–156); thus, offering a more truthful view of the city.

Finally, the research process managed to incorporate all three phases of transdisciplinarity outlined by Lang et al. (2012:27). The first phase is concerned with collaboratively framing the problem and building a collaborative research team. This was achieved through multiple project meetings and the selection and inclusion of four diverse communities and NGOs working with them. The second phase involves co-producing solution-orientated and transferable knowledge through collaborative research. The 'research' for the regenerative public space project took place through multiple large and small workshops aimed at creating a shared knowledge based on regenerative space and then exploring the detailed process in each specific context with the particular stakeholders. The third phase focuses on the (re)integration and application of the produced knowledge in both scientific and societal practise. This will occur through the two workshops planned for 2024 and the digital platform to ensure ongoing interaction, integration and application in the future.

An added benefit is that the phases of transdisciplinarity also open up the possibility for the ongoing process of regenerative development and design, which is concerned with the co-evolution of humans and nature, and designed to be 'tested' and consciously applied through the digital platform. Mang and Reed (2012) offer guidance on the process to initiate and facilitate regenerative development and design. This requires a focus on three phases, namely, (1) understanding the place, (2) aligning place and people, and (3) ensuring ongoing care through co-mutualism. Facilitating a process of regenerative development requires an in-depth *understanding of the place* as a way for people to envision the unity of humans and nature in a concrete and specific way, identify specific patterns and understanding the patterns and relationships in order to reveal the living qualities of a site and its place. Finding and narrating the story creates an opportunity for potential to be revealed, which is the unique essence of the place and people (Mang & Reed, 2012; Mang et al., 2016). Through these stakeholder workshops, focus groups in the community and site analysis while walking in space and discussing the spatial elements the project realised a form of action research.

This establishes a foundation for the second phase—namely, *to design for harmony with the place*. Such a design should aim to harmonise the buildings and infrastructure with the patterns of the land and its ecological and cultural systems towards improvement of the whole system (Mang & Reed, 2012; Mang et al., 2016; Hes & Du Plessis, 2014). Harmonising buildings and infrastructure with the ecological and cultural systems means that the interventions and the site are considered as a living system with its own unique characteristics. Effective solutions cannot be based on formulas or be copied from somewhere else (Hes & Du Plessis, 2014).

Finally, Regenerative Development and Design is not complete with the final drawings, approvals or construction of the product. It also requires that the regenerative capacity of the project and the people who inhabit and manage it be sustained—in other words, that it catalyses a process of *co-evolving mutualism*. As a project cannot supply the resources and energy to achieve the full potential of its ongoing regenerative effect, it must call into existence a system of mutually beneficial stakeholder relationships to organise cooperative enterprises required to enable evolution. Co-evolving mutualism cannot be predicted, but it can be continually planned and managed (Mang & Reed, 2012; Mang et al., 2016). The project established the preconditions for these phases through the development of a digital platform to facilitate continuous engagement and emergence of opportunities over time.

The combination of these three factors of transdisciplinarity in the project and its overlap with the process of regenerative sustainability is illustrated in Fig. 8.3. This allows for the blurring of boundaries through collaboration, participation and cooperation.

8.5 Conclusion

This chapter discussed a case study of the City of Tshwane in South Africa, featuring a project on regenerative public space. The project deals with four different settlement areas in the municipality and their various challenges and potential for regenerative public space. The case study highlighted various methods concerned with the spatiality of the city utilised through a transdisciplinarity approach for urban sustainability. These methods and processes are used to develop a methodology for regenerative public space and build a digital platform to contribute to the implementation of the process in practise through online interaction and knowledge sharing. The case study showed that research methods applied to spatial phenomenon for regenerative urban sustainability and regenerative public space are concerned with three actions—collaboration, participation and cooperation—through transdisciplinarity across disciplines and beyond academia. Regenerative sustainability can be enhanced through understanding, adapting and regenerating urban space through an integration of three structural elements—function, form and flows in spatial planning and design interventions. To do this in a holistic way requires a shared conceptual framework and a set of spatially orientated methods to offer a holistic symbiosis of the whole system.

The development of a digital platform assists to facilitate an ongoing process of co-evolution through (1) access to data and information to highlight potential within the four communities and allow others to learn from it, (2) taking research activities outside the university to accommodate co-production of knowledge and facilitate participation, and (3) building ongoing cooperation between communities, academia and professionals from private practice towards greater knowledge sharing and integration. The RPS digital platform symbolises new modes of collaborative project planning and research potential between multidisciplinary teams, with technology as the medium and the Internet as a catalyst. The ability to craft GIS-enabled digital platforms that can combine sector-specific content, workflows, resources and outputs across an array of specialisations via the Internet ushers in a new era of transdisciplinary collaboration. Dealing with socio-ecological systems across scales in different contexts necessitates a transdisciplinary approach to address urban sustainability and facilitate ongoing transformative process in practice, for both the environment and the people involved in the process.

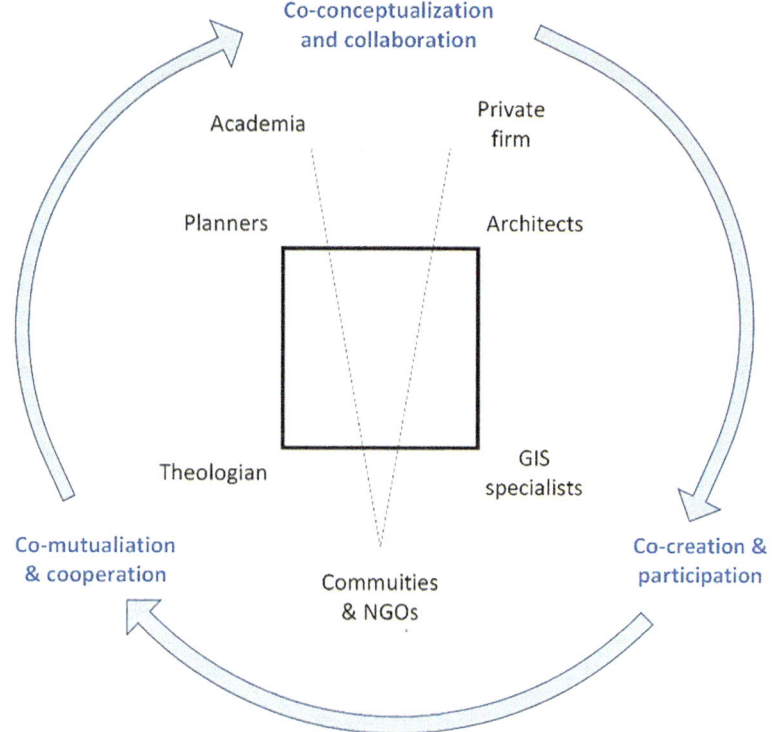

Fig. 8.3 A transdisciplinary approach involving academia and practise through collaboration, participation and cooperation through regenerative development and design. (Source: authors)

References

Balsiger, P. W. (2004). Supradisciplinary research practices: History, objectives and rationale. *Futures, 36*, 407–421.

Balsiger, J. (2015). Transdiciplinarity in the class room? Simulating co-production of sustainability knowledge. *Futures, 65*, 185–194.

Benne, B., & Mang, P. (2015). Working regeneratively across scales—Insights from nature applied to the built environment. *Journal of Cleaner Production, 109*, 42–52.

Bruce, A., Lyall, C., Tait, J., & Williams, R. (2004). Interdisciplinary integration in Europe: The case of the Fith framework programme. *Futures, 36*, 357–470.

Capra, F. (2016). *The web of life*. Harper-Collins Publishers.

Cupers, K., Oldfield, S., Herz, M., Nkula-Wenz, L., Disretti, E., & Perret, M. (2022). *What is critical urbanism? Urban research as pedagogy*. Park Books.

Doucet, I., & Janssens, N. (2011). Editorial: Transdisciplinarity, the hybridization of knowledge production and space-related research. In I. Doucet & N. Janssens (Eds.), *Transdisciplinary knowledge production in architecture and urbanism: Towards hybrid modes of inquiry* (pp. 1–14). Springer.

Du Plessis, C. (2012). Towards a regenerative paradigm for the built environment. *Building Research and Information, 40*(1), 7–22.

Du Plessis, C. (2022). The city sustainable, resilient, regenerative—A rose by any other name? In R. Roggema (Ed.), *Design for regeneration*. Springer.

Franklin, A., & Blyton, P. (2011). Sustainability research: An introduction. In A. Franklin & P. Blyton (Eds.), *Researching sustainability: A guide to social science methods, practice and engagement* (pp. 3–16). Earthscan.

Gibbons, L. V. (2020). Regenerative—The new sustainable? *Sustainability, 12*, 5483.

Hes, D., & Du Plessis, C. (2014). *Designing for Hope: Pathways to regenerative sustainability*. Earthscan for Routledge.

Hopwood, B., & Mellor, M. (2005). Sustainable development: Mapping different approaches. *Sustainable Development, 13*(1), 38–52. https://doi.org/10.1002/sd.244

Klein, J. (1990). *Interdisciplinarity: History, theory and practice*. Wayne State University Press.

Lang, D. J., Wiek, A., Bergmann, M., Stauffacher, M., Martens, P., Moll, P., Swilling, M., & Thomas, C. J. (2012). Transdisciplinary research in sustainability science: Practice, principles, and challenges. *Sustainability Science, 7*(Suppl 1), 25–43.

Lawrence, R. J. (2004). Housing and health: From interdisciplinary principles to transdisciplinary research and practice. *Futures, 36*, 487–502.

Leemans, R. (2016). The lessons learned from shifting from global-change research programmes to transdisciplinary sustainability science. *Current Opinion in Environmental Sustainability, 19*, 103–110.

Mang, P., & Reed, B. (2012). Designing from place: A regenerative framework and methodology. *Building Research & Information, 40*(1), 23–38.

Mang, P., Haggard, B., & Regenesis. (2016). *Regenerative development and design. A framework for evolving sustainability*. John Wiley & Sons Ltd.

Marsden, T. (2011). Sustainability science and a new spatial imagination: Exploring some Analytical and methodological considerations. In A. Franklin & P. Blyton (Eds.), *Researching sustainability: A guide to social science methods, practice and engagement* (pp. 297–316). Earthscan.

Polk, M. (2015). Transdiciplinary co-production: Designing and testing a transdisciplinary research framework for societal problem solving. *Futures, 65*, 110–122.

Ramadier, T. (2004). Transdisciplinarity and its challenges: The case of urban studies. *Futures, 36*, 423–439.

Reed, B. (2007). Shifting from 'sustainability' to regeneration. *Building Research & Information, 35*(6), 674–680.

Schauppenlehner-Kloyber, E., & Penker, M. (2014). Managing group processes in transdisciplinary future studies: How to facilitate social learning and capacity building for self-organised action towards sustainable urban development? *Futures, 65*, 57–71.

World Commission on Environment and Development (WCED) (1987). Bruntland report, United Nations.

Karina Landman is a Professor and HOD in the Department of Town and Regional Planning at the University of Pretoria with a background in Architecture and Urban Design. Her work focuses on sustainable development, including resilience and regenerative sustainability and public space.

Ilan Guest is a professional town planner and Geographic Information Systems (GIS) specialist focused on spatial policy, rural development and online GIS applications. Guest concluded his MA in City and Regional Planning at the Georgia Institute of Technology in Atlanta Georgia (USA) as a Fulbright Scholar with specialisation in GIS and Remote Sensing. He is also an external lecturer at the Town Planning Department at the University Pretoria, where he teaches topics around regional and rural development, and the role of GIS and spatial data applications for the planning profession.

Open Access This chapter is licensed under the terms of the Creative Commons Attribution 4.0 International License (http://creativecommons.org/licenses/by/4.0/), which permits use, sharing, adaptation, distribution and reproduction in any medium or format, as long as you give appropriate credit to the original author(s) and the source, provide a link to the Creative Commons license and indicate if changes were made.

The images or other third party material in this chapter are included in the chapter's Creative Commons license, unless indicated otherwise in a credit line to the material. If material is not included in the chapter's Creative Commons license and your intended use is not permitted by statutory regulation or exceeds the permitted use, you will need to obtain permission directly from the copyright holder.

Conclusion: Recommendations for Spatial-Methodological, Transdisciplinary Action Regarding SDG11 (Position Paper)

Fraya Frehse, Angela Million, Ignacio Castillo Ulloa, Jacques du Toit, Amy Pieterse, Jenia Mukherjee, Shreyashi Bhattacharya, Caio Moraes Reis, Giulia Pereira Patitucci, Budi Heru Santosa, Raldi Hendro Koestoer, and Inês Martina Lersch

The truly southern-global spatial-methodological, transdisciplinary endeavour, which we have gone through in the previous chapters, led us to cities as diverse as Durban and Tshwane (South Africa), San Jose (Costa Rica), Porto Alegre and São Paulo (Brazil), Kolkata (India), Bangkok (Thailand) and Tangerang (Indonesia). Both academics and practitioners of diverse professional and personal backgrounds got together in each chapter to deploy spatial methods jointly, in transdisciplinarity, for the purpose of various targets related to SDG11—ranging from inclusion,

F. Frehse (✉)
Department of Sociology, University of São Paulo + SMUS, São Paulo, SP, Brazil
e-mail: fraya@usp.br

A. Million
Department of Urban and Regional Planning + SMUS, Technische Universität, Berlin, Germany

I. Castillo Ulloa
Department of Urban and Regional Planning + SMUS, Technische University Berlin, Berlin, Germany

J. du Toit
Department of Town and Regional Planning, University of Pretoria, Pretoria, South Africa

A. Pieterse
Smart Places Cluster, Council for Scientific and Industrial Research, Pretoria, South Africa

J. Mukherjee
Department of Humanities and Social Sciences, Indian Institute of Technology Kharagpur, Kharagpur, West Bengal, India

S. Bhattacharya
Senior Research Fellow, Rekhi Centre of Excellence for the Science of Happiness, Indian Institute of Technology, Kharagpur, West Bengal, India

C. M. Reis
Department of Sociology + SMUS, University of São Paulo, São Paulo, SP, Brazil

G. P. Patitucci
Federal Ministry of Management and Innovation in Public Services, Brasília, Brazil

B. H. Santosa
National Research and Innovation Agency (BRIN), Jakarta, Indonesia

R. H. Koestoer
School of Environmental Science, Universitas Indonesia, Jakarta, Indonesia

I. M. Lersch
Department of Urban Planning + SMUS, Federal University of Rio Grande do Sul, Porto Alegre, Brazil

© The Editor(s) (if applicable) and The Author(s) 2025
F. Frehse et al. (eds.), *Spatial Methods in Transdisciplinarity for Urban Sustainability*, Sustainable Development Goals Series, https://doi.org/10.1007/978-3-031-84367-9

resilience and sustainability in public spaces, to urban heritage and environmental risks.

In the wake of this diversified set of methodological experiences and reflections, we are well aware that the academic, professional and personal lessons learned are uncountable. These most likely transcend the limits of any concluding chapter. Things change, however, once we turn the opportunity of a conclusion into a position paper of a spatial-methodological nature regarding transdisciplinary projects towards SDG11. What follows is therefore a set of recommendations for spatial-methodological, transdisciplinary action towards SDG11.

The Transformative Role of Spatial Methods

By pursuing very diversified research-and-practice paths with the aid of specific combination of spatial methods, each of the eight chapters has contributed in peculiar ways to make evident the book's central claim: the deployment of spatial methods in transdisciplinary projects for SDG11 has a *transformative* role. While some chapters especially highlight the *personal* dimension of the changes brought about to academics by the spatial-methodological, transdisciplinary experiments (Chaps. 2, 4, and 5), others (Chaps. 1, 3, 6, 7, and 8) emphasised the *academic* reach of the spatial-methodological experience accomplished in and through transdisciplinarity.

The substantially transformative essence of SDG11 comes as no surprise once we reflect upon the methodological implications of this goal for projects that address it. Transdisciplinarity is highly welcome when the aim is to *practically* contribute to making cities and urban settlements 'inclusive, safe, resilient and sustainable' by means of *evidence-based knowledge*. The book chapters demonstrate that the social and cultural complexity of current cities and urban settlements almost inherently challenges academics and practitioners to co-working. Even more so when it comes to the socially widest possible production and dissemination of evidence-based knowledge.

SDG11 goes hand in hand with transdisciplinary projects, which bring together stakeholders from various professional backgrounds, as the previous pages well showed. We are referring to people who specialise both in producing and transferring academic knowledge, and experts (of various ages!) in the production and dissemination of other, likewise crucial social forms of knowledge, ranging from ecology and activism to the arts and politics, and vice-versa.

Thus, we reach the second decisive methodological implication of SDG11, which the book chapters make explicit. If SDG11 encourages transdisciplinary projects, these necessarily are *spatially sensitive*. Not only are cities and urban settlements spaces in their own right and make evident that space is a social product—that is, a set of bodily and materially mediated relations between human beings and social/symbolic objects, which has been conceptualised in various ways since Lefebvre's pioneering approach. Due to this social nature of space, transdisciplinarity is therefore essentially spatial. The co-production of evidence-based knowledge in the framework of research-and-practice collaboration between scientific researchers and local practitioners (based in NGOs, private firms or government agencies) as well as independent policy makers or artists only happens because it literally 'takes place'; that is, it simultaneously is spatial and produces space(s).

Hence, the third decisive methodological dimension of SDG11-focused projects: whether explicitly or implicitly, they make use of *spatially sensitive methods*. The majority of the book author teams (Chaps. 1, 2, 3, 4, 5, 6, and 8) explicitly stressed the pertinence of spatial methods—that is, sets of empirical research techniques sensitive to the social dimension of urban spaces. Elsewhere (Chap. 7), the pertinence of a spatial-methodological toolkit was connected to improvements in understanding and policy regarding the physical materiality of urban space. Well aware that these different approaches are epistemologically inseparable from their underlying, theoretically based concept of space, the crucial aspect here lies in method rather than theory. Altogether, the book chapters make evident that transdisciplinary projects for SDG11 cannot do it without spatially sensitive methods. Accordingly, transdisciplinary projects for SDG11 offer spatially, hence empirically, visible clues about the qualitative and transformative dimension implicit in the research-and-practice process of deploying spatial methods precisely in transdisciplinarity for SDG11.

In light of the methodological implications above, our first recommendation would be that *stakeholders of transdisciplinary projects for SDG11 should pay critical attention to the personally and politically transformative role that spatial methods may play in their projects.*

This said, the transformations themselves via spatial methods become an issue of interest—and of recommendation. Altogether, the book chapters demonstrate transformative possibilities and limitations of spatial methods in transdisciplinarity for urban sustainability.

The Combination of Spatial Methods

As to the potential of spatial methods, their critical reflection and/or deployment throughout the chapters has led us to acknowledge the transformative possibilities implicit in *combining* various spatial methods in transdisciplinary projects for SDG11. But the combination at stake is not random. It *depends on the respective research-and-practice problem*, or major question, addressed by each project. Furthermore, it often implies methodological *articulations between historically quantitative designs and tools* such as GIS systems (Chaps. 1, 2, 7, and 8), (participatory) mapping (Chaps. 1, 2, 3, 4, and 5) and surveys (Chap. 7), and *qualitative designs and tools* such as techniques of ethnographic observation and visualisation (Chaps. 4, 5, and 6) apart from interviews and their variations (Chaps. 3 and 7), as well as explicitly participatory approaches (Chaps. 3 and 8).

Additionally, it is through such strategic combinations that the further development of existing spatial methods occurs (Chaps. 3, 4, and 5). When diverse methods are brought together to address complex urban challenges, they not only complement each other but also evolve to meet the specific needs of the project. This enhancement is particularly evident in how traditional methods are adapted and refined in response to new contexts and challenges encountered in the field. Moreover, these combinations can lead to the emergence of entirely new spatial methods (Chap. 4). The need for new and adaptable methods became particularly clear during the global COVID-19 pandemic (Chaps. 3 and 4), which introduced unprecedented challenges to spatial research. The restrictions on mobility and social interaction required researchers to innovate rapidly, rethinking the use of spatial techniques that could operate effectively under these constraints. These innovations, though initially born out of necessity, have since contributed to the broader advancement of spatial methods within transdisciplinary research.

To be sure, our second recommendation is that *this multi-dimensional use of spatial methods for SDG11—that is, their (i) combination, further and new development (ii) forged in research-and-practice questions (iii) with the aid of quantitative and qualitative techniques—follows social-learning aims* (Chap. 2). In other words, the very use of spatial methods as technical tools of application in transdisciplinary projects for SDG11 is enhanced by means of the stakeholders' outputs, their traditional skills and forms of local knowledge framed by the specific local contexts and global circumstances they navigate.

Three Possibilities Implicit in Spatial Methods

By focusing on the contributions of spatial methods to urban planning research (Chap. 1), practice (Chaps. 2, 7, and 8) and teaching (Chap. 3), five book chapters showed that transdisciplinary urban planning for SDG11 is enhanced when *urban planners carefully reflect about spatial methods throughout their research*. Issues of particular interest are the empirical applicability of spatial methods as well as their epistemic and social impacts in local contexts. Even more so when it comes to integrating participatory processes within the projects in focus (Chaps. 3, 7, and 8). Conflicts over urban development projects are inevitable (Chap. 3) vis-à-vis the variably distinct logics implicit in research and practice (Chaps 1 and 2).

Put differently, in the book's social-scientifically inspired chapters (Chaps. 4, 5, and 6), methodological ponderings appear as intrinsic elements of each and every research-and-practice project. Therefore, for social scientists, the novelty of spatial methods lies somewhere else. It consists of the *joint social-learning impact of the ethnographic method for addressing SDG11 in a transdisciplinary way*. By, respectively, exploring the bodily and materially interactional (Chap. 4), the visual (Chap. 5) and the multispecies (Chap. 6) facets of ethnography to accomplish transdisciplinarity in practice, the author teams highlighted that the social learning process triggered by the project's accomplishment as such concerns both academic and practitioners with diverse backgrounds. Their protagonists were not only junior and senior academics and practitioners ranging from social workers to ex-homeless people—but adolescent school students as well!

Considering the forementioned, two possibilities implicit in the deployment of spatial methods in the book are due to the inter- and transdisciplinary differences among and between the chapter author teams, while a third spatial-methodological possibility is based on an unexpected similarity among them. In their various empirical ways of deploying spatial methods, the chapters make evident that spatial methods are vigorous, social communication devices in transdisciplinary projects for SDG11. They help bridging the understanding, reciprocal empathy—and knowledge co-production—among academics and practitioners. It is crucial, however, that academics are willing and able to collaborate closely with practitioners, encouraging them to recognise that their knowledge is a crucial part of transdisciplinary research. Yet, the challenge of involving practitioners in co-writing and publishing transdisciplinary findings remains, as was also noted in some chapters. This suggests the need for new ways of co-authoring or exploring different formats of dissemination.

To sum up, our third recommendation concerns *the threefold possibilities of spatial methods in transdisciplinarity for SDG11: (i) there is the need to move from implicit reliance to a more reflective and explicit use of spatial methods in urban planning research, practice and teaching; (ii) the social learning impact of ethnographic spatial methods must be recognised and leveraged to enhance transdisciplinary collaboration; (iii) spatial methods can be effectively utilised in their social communication role in transdisciplinarity.*

Four Limitations of Spatial Methods

The first limitation is precisely of a methodological nature. The deployment of spatial methods presupposes technical skills in research which are by no means generalisable among practitioners and communities—even more so in socially unequal countries such as the ones addressed in all the chapters. Teaching curricula in disciplines, apart from the social sciences, often have limited and insufficient methodological training in quantitative and qualitative research.

A second limitation is the intrinsically qualitative nature of the social learning process implicit in the transdisciplinary deployment of spatial methods towards SDG11 (Chaps. 2, 4, and 5). The assessment of post-dissemination outcomes or outputs among practitioners is an essentially qualitative procedure, which demands time, space and funding.

The third limitation concerns transdisciplinary participation and engagement. Following successful spatial-methodological transdisciplinary projects, there is always the challenge of integrating participation within subsequent projects (Chap. 2). The challenges of the ivory tower apart from time and budget (Chap. 7) are always there. Not to mention the socio-economic, cultural and geographical conditions of each city to be addressed with the aid of a specific set of spatial methods (Chap. 8).

A fourth limitation is that spatial methods and transdisciplinarity are risky. The intended outcomes or impacts are not guaranteed; and even if so, how do we really measure such outcomes and impacts? Transdisciplinary projects can have unintended outcomes (Chaps. 1 and 4), good and bad. This is, respectively, due to their integrated nature, to the politically, and socially diverse groups involved, and to the fact that they take place in an empirically given, 'real' world context, and hence deal with 'real' world dynamics. Therefore, reflexivity and flexibility are not only essential when developing or designing a project. The same applies to setting up frameworks and measures to evaluate the project's outcomes and impacts. Such evaluations also need to consider both societal impacts and scientific impacts. And this adds another layer of complexity to the matter.

Against the backdrop of the social and political width of the dilemmas implicit in the four forementioned limitations, we would like to finish this position paper with one fourth recommendation: *Spatial-methodologically supported projects in transdisciplinarity for SDG11 should prioritise the formation of research-and-practice teams that* include *(i) members with strong spatial-methodological skills, (ii) individuals who can bridge the gap between technical expertise and local knowledge, (iii) and experts in participatory processes who can ensure that stakeholder and community engagement is meaningful and sustained.*

CONCLUSIÓN: Recomendaciones para una acción espacio-metodológica transdisciplinaria en relación con el ODS11 (Documento de posición)

Fraya Frehse, Angela Million, Ignacio Castillo Ulloa, Jacques du Toit, Amy Pieterse, Jenia Mukherjee, Shreyashi Bhattacharya, Caio Moraes Reis, Giulia Pereira Patitucci, Budi Heru Santosa, Raldi Hendro Koestoer, and Inês Martina Lersch

La labor espacio-metodológica y transdisciplinaria del sur del globo que hemos llevado a cabo a lo largo de los capítulos anteriores nos ha trasladado a ciudades tan diversas como Durban y Tshwane (Sudáfrica), San José (Costa Rica), Porto Alegre y São Paulo (Brasil), Calcuta (India), Bangkok (Tailandia) y Tangerang (Indonesia). Tanto académicos como profesionales de diferentes ámbitos profesionales y personales han colaborado en cada capítulo para

F. Frehse · C. M. Reis
Department of Sociology + SMUS, University of São Paulo, São Paulo, SP, Brazil
e-mail: fraya@usp.br

A. Million
Department of Urban and Regional Planning + SMUS, Technische Universität, Berlin, Germany

I. Castillo Ulloa
Department of Urban and Regional Planning + SMUS, Technische Universität Berlin, Berlin, Germany

J. du Toit
Department of Town and Regional Planning, University of Pretoria, Pretoria, South Africa

A. Pieterse
Smart Places Cluster, Council for Scientific and Industrial Research, Pretoria, South Africa

J. Mukherjee
Department of Humanities and Social Sciences, Indian Institute of Technology Kharagpur, Kharagpur, West Bengal, India

S. Bhattacharya
Senior Research Fellow, Rekhi Centre of Excellence for the Science of Happiness, Indian Institute of Technology, Kharagpur, West Bengal, India

G. P. Patitucci
Federal Ministry of Management and Innovation in Public Services, Brasília, Brazil

B. H. Santosa
National Research and Innovation Agency (BRIN), Jakarta, Indonesia

R. H. Koestoer
School of Environmental Science, Universitas Indonesia, Jakarta, Indonesia

I. M. Lersch
Department of Urban Planning + SMUS, Federal University of Rio Grande do Sul, Porto Alegre, Brazil

desplegar métodos espaciales de forma conjunta y con transdisciplinariedad para conseguir varios objetivos relacionados con el ODS11, que van desde la inclusión, la resiliencia y la sostenibilidad en espacios públicos, hasta el patrimonio urbano y los riesgos medioambientales.

A raíz de este conjunto diversificado de experiencias y reflexiones metodológicas, somos muy conscientes de que las lecciones académicas, profesionales y personales aprendidas son innumerables. Es muy probable que estas trasciendan los límites de cualquier capítulo de conclusiones. No obstante, las cosas cambian una vez que convertimos la oportunidad de una conclusión en un documento de posición de naturaleza espacio-metodológica en relación con los proyectos transdisciplinarios orientados al ODS11. Por lo tanto, lo que sigue es un conjunto de recomendaciones para una acción transdisciplinaria de carácter espacio-metodológico orientada al ODS11.

El papel transformador de los métodos espaciales

Al buscar rutas de investigación y práctica muy diversificadas con la ayuda de una combinación específica de métodos espaciales, cada uno de los ocho capítulos ha contribuido de manera diferente a evidenciar el argumento principal del libro: el despliegue de métodos espaciales en proyectos transdisciplinarios para el ODS11 tiene un papel *transformador*. Mientras que algunos capítulos destacan especialmente la dimensión *personal* de los cambios provocados en el mundo académico a través de experimentos espacio-metodológicos y transdisciplinarios (capítulos 2, 4, 5), otros (capítulos 1, 3, 6, 7 y 8) enfatizan el alcance *académico* de la experiencia espacio-metodológica lograda en y a través de la transdisciplinariedad.

La esencia sustancialmente transformadora del ODS11 no resulta extraña una vez que reflexionamos sobre las implicaciones metodológicas de este objetivo para los proyectos orientados a ello. La transdisciplinariedad es muy bien recibida cuando el objetivo es contribuir de *forma práctica* a que las ciudades y núcleos urbanos sean «inclusivos, seguros, resilientes y sostenibles» mediante un *conocimiento basado en pruebas*. Los capítulos del libro demuestran que la complejidad social y cultural de las ciudades y núcleos urbanos actuales supone un reto casi inherente para la colaboración entre académicos y profesionales. Más aún cuando se trata de una producción y difusión lo más amplia posible desde el punto de vista social de los conocimientos basados en pruebas. El ODS11 va mano a mano con proyectos transdisciplinarios que reúnen a partes interesadas de diferentes ámbitos profesionales como bien se ha mostrado en páginas anteriores. Nos referimos tanto a personas especializadas en la producción y transmisión de conocimientos académicos, como a expertos (de diferentes edades) en producción y difusión de otras formas de conocimiento social igualmente cruciales, que van desde la ecología y el activismo hasta el arte y la política, y viceversa.

De esta manera, llegamos a la segunda implicación metodológica decisiva del ODS11, que ponen de manifiesto los capítulos del libro. Si el ODS11 promueve proyectos transdisciplinarios, estos necesariamente son *espacialmente sensibles*. Las ciudades y núcleos urbanos no solo son espacios por derecho propio y ponen de manifiesto que el espacio es un producto social, es decir, un conjunto de relaciones físicas y materialmente mediadas entre los seres humanos y los objetos sociales/simbólicos que se ha conceptualizado de varias maneras desde el enfoque pionero de Lefebvre. Debido a esta naturaleza social del espacio, la transdisciplinariedad es esencialmente espacial. La coproducción de conocimientos basados en pruebas en el marco de la colaboración de investigación y práctica entre investigadores científicos y profesionales locales (ONG, empresas privadas y agencias gubernamentales), así como los responsables políticos o artistas independientes, solo ocurre porque literalmente «pasa»; es decir, es simultáneamente espacial y produce espacio(s).

Por lo tanto, la tercera dimensión metodológica decisiva del ODS11 se enfoca en proyectos, explícita o implícitamente, que usan *métodos espacialmente sensibles*. La mayoría de los equipos de autores del libro (capítulos 1, 2, 3, 4, 5, 6 y 8) subrayaron explícitamente la idoneidad de los métodos espaciales, es decir, los conjuntos de técnicas de investigación empírica sensibles a la dimensión social de los espacios urbanos. En otra parte (el capítulo 7), la importancia de un kit de herramientas espacio-metodológicas se vinculó a mejoras en el entendimiento y las políticas considerando la materialidad física del espacio urbano. Siendo muy conscientes de que estos enfoques diferentes son epistemológicamente inseparables de su concepto subyacente del espacio, basado en la teoría, el aspecto crucial radica aquí más en el método que en la teoría. En general, los capítulos del libro evidencian que los proyectos transdisciplinarios para el ODS11 no pueden realizarse sin métodos espacialmente sensibles. En consecuencia, los proyectos transdisciplinarios para el ODS11 ofrecen claves visibles desde el punto de vista espacial y, por lo tanto, empíricamente, sobre la dimensión cualitativa y transformadora implícita en el proceso de investigación y práctica del despliegue de métodos espaciales precisamente en la transdisciplinariedad para el ODS11.

A la luz de las implicaciones metodológicas anteriores, nuestra primera recomendación sería que *las partes interesadas en proyectos transdisciplinarios para el ODS11 deberían prestar especial atención al papel personal y políticamente transformador que los métodos espaciales pueden desempeñar en sus proyectos.*

Dicho esto, las transformaciones en sí mismas a través de los métodos espaciales se convierten en una cuestión de interés y de recomendación. En resumen, los capítulos del libro demuestran las posibilidades transformadoras y las limitaciones de los métodos espaciales en la transdisciplinariedad para la sostenibilidad urbana.

La combinación de métodos espaciales

En cuanto al potencial de los métodos espaciales, su reflexión crítica y/o despliegue a lo largo de los capítulos nos ha llevado a reconocer las posibilidades transformadoras implícitas en la *combinación* de diferentes métodos espaciales en proyectos transdisciplinarios para el ODS11. Sin embargo, la combinación en cuestión no es aleatoria: *depende del respectivo problema de investigación y práctica* o asunto principal, que aborde cada proyecto. Además, con frecuencia implican *articulaciones metodológicas entre diseños y herramientas históricamente cuantitativas* como los sistemas SIG (capítulos 1, 2, 7 y 8), la cartografía (participativo) (capítulos 1, 2, 3, 4 y 5), las encuestas (capítulo 7), los *diseños y herramientas cualitativas* como técnicas de observación y visualización etnográficas (capítulos 4, 5 y 6), las entrevistas y sus variaciones (capítulos 3 y 7), y los enfoques explícitamente participativos (capítulos 3 y 8).

Además, a través de dichas combinaciones estratégicas es como se produce un mayor desarrollo de los métodos espaciales existentes (capítulos 3, 4 y 5). Cuando se reúnen diversos métodos para abordar retos urbanos complejos, no solo se complementan entre sí, sino que evolucionan para satisfacer las necesidades específicas del proyecto. Esta mejora es particularmente evidente en el modo en que los métodos tradicionales se adaptan y perfeccionan como respuesta a las nuevas situaciones y retos que surgen en este campo. Asimismo, estas combinaciones pueden dar lugar a métodos espaciales completamente nuevos (capítulo 4). La necesidad de métodos nuevos y adaptables se hizo claramente evidente durante la pandemia de la COVID-19 (capítulos 3 y 4), ya que trajo desafíos sin precedentes para la investigación espacial. Las restricciones de movilidad e interacción social obligaron a los investigadores a innovar con rapidez, replanteándose el uso de técnicas espaciales que pudieran funcionar eficazmente bajo estas limitaciones. Dichas innovaciones, aunque nacieran inicialmente fruto de la necesidad, han contribuido desde entonces a un mayor avance de los métodos espaciales dentro de la investigación transdisciplinaria.

Sin duda, nuestra segunda recomendación es que *este uso multidimensional de los métodos espaciales para el ODS11, es decir, su (i) combinación, profundización y nuevo desarrollo (ii) forjado en cuestiones de investigación y práctica*

(iii) con la ayuda de técnicas cuantitativas y cualitativas, siga los objetivos de aprendizaje social (capítulo 2). Dicho de otro modo, el mero uso de métodos espaciales como herramientas técnicas de aplicación en proyectos transdisciplinarios para el ODS11 se ve fortalecido por medio de las aportaciones de las partes interesadas, sus habilidades tradicionales y las formas de conocimiento locales dentro de un contexto local específico y circunstancias globales por las que transitan.

Tres posibilidades implícitas en los métodos espaciales

Poniendo el foco en las contribuciones de los métodos espaciales a la investigación (capítulo 1), a la práctica (capítulos 2, 7 y 8) y a la enseñanza (capítulo 3) urbanística, cinco capítulos del libro muestran que la planificación urbana transdisciplinaria para el ODS11 mejora cuando los *planificadores urbanísticos reflexionan detenidamente sobre los métodos espaciales durante toda su investigación.* Las cuestiones de especial interés son la aplicabilidad empírica de métodos espaciales, así como sus repercusiones epistémicas y sociales en contextos locales, especialmente ún cuando se trata de integrar procesos participativos en los proyectos en cuestión (capítulos 3, 7 y 8). Los conflictos en torno a los proyectos de desarrollo urbanístico son inevitables (capítulo 3) con respecto a las distintas lógicas variables implícitas en la investigación y la práctica (capítulos 1 y 2).

En otras palabras, en los capítulos de inspiración científico-social del libro (capítulos 4, 5 y 6), las reflexiones metodológicas aparecen como elementos intrínsecos de cada uno de los proyectos de investigación y práctica. Por lo tanto, para los científicos sociales, la novedad de los métodos espaciales reside en otra parte: esta consiste en el *impacto conjunto de aprendizaje social del método etnográfico para abordar el ODS11 de un modo transdisciplinario.* Al explorar respectivamente las facetas física y materialmente interactiva (capítulo 4), visual (capítulo 5) y multiespecífica (capítulo 6) de la etnografía para

conseguir la transdisciplinariedad en la práctica, el equipo de autores recalcó que el proceso de aprendizaje social desencadenado por la realización del proyecto como tal concierne tanto a académicos como a profesionales de diferentes ámbitos. Sus protagonistas no solo eran académicos y profesionales júnior y sénior: ¡iban desde trabajadores sociales hasta personas sin hogar y también estudiantes adolescentes!

Considerando lo mencionado anteriormente, dos posibilidades implícitas en el despliegue de métodos espaciales en el libro se deben a las diferencias interdisciplinarias y transdisciplinarias entre los equipos de autores de los capítulos, mientras que una tercera posibilidad espacio-metodológica se basa en una similitud inesperada entre ellos. En sus diferentes formas empíricas de despliegue de métodos espaciales, los capítulos ponen de manifiesto que los métodos espaciales son dispositivos de comunicación social fuertes en los proyectos transdisciplinarios para el ODS11. Estos ayudan a tender puentes de entendimiento, empatía recíproca y coproducción de conocimientos entre académicos y profesionales. No obstante, es determinante que los académicos estén dispuestos y sean capaces de colaborar estrechamente con los profesionales, animándoles a reconocer que sus conocimientos son una parte crucial de la investigación transdisciplinaria. Sin embargo, queda el reto de implicar a los profesionales en la redacción conjunta y la publicación de los resultados transdisciplinarios, como se señala en algunos capítulos. Esto sugiere la necesidad de nuevas maneras de coautoría o explorar diferentes formatos de difusión.

En resumen, nuestra tercera recomendación se refiere a *las tres posibilidades de los métodos espaciales en la transdisciplinariedad para el ODS11: (i) es necesario pasar de la·confianza implícita a un uso más reflexivo y explícito de los métodos espaciales en la investigación, práctica y enseñanza urbanística; (ii) el impacto del aprendizaje social de los métodos espaciales etnográficos debe reconocerse y aprovecharse para fortalecer la colaboración transdisciplinaria; (iii) los métodos espaciales pueden utilizarse de manera efectiva en su función de comunicación social en la transdisciplinariedad.*

Cuatro limitaciones de los métodos espaciales

La primera limitación es precisamente de naturaleza metodológica. El despliegue de métodos espaciales presupone habilidades técnicas en investigación que no pueden generalizarse en absoluto entre los profesionales y las comunidades, más aún en países con desigualdades sociales como los que se abordan en todos los capítulos. Los planes de enseñanza de las disciplinas, además de las ciencias sociales, suelen tener una formación metodológica limitada e insuficiente en investigación cuantitativa y cualitativa.

Una segunda limitación es la naturaleza intrínsecamente cualitativa del proceso de aprendizaje social implícito en el despliegue transdisciplinario de los métodos espaciales orientados al ODS11 (capítulos 2, 4 y 5). La evaluación de los resultados o productos posteriores a la difusión entre profesionales es un procedimiento fundamentalmente cualitativo que requiere tiempo, espacio y financiación.

La tercera limitación es la participación y el compromiso transdisciplinarios. Tras el éxito de los proyectos transdisciplinarios de metodología espacial, siempre quedará el reto de integrar la participación en proyectos futuros (capítulo 2). Los desafíos de la torre de marfil, aparte del tiempo y el presupuesto (capítulo 7), siempre están presentes. Por no mencionar las condiciones socioeconómicas, culturales y geográficas de cada ciudad que se abordan con la ayuda de un conjunto específico de métodos espaciales (capítulo 8).

Una cuarta limitación es que los métodos espaciales y la transdisciplinariedad entrañan riesgos. Los resultados o impactos previstos no están garantizados y, si así fuera, ¿cómo mediríamos realmente dichos resultados e impactos? Los proyectos transdisciplinarios pueden tener resultados imprevistos (capítulos 1 y 4), tanto buenos como malos. Esto se debe, respectivamente, a su naturaleza integrada, a los diferentes grupos sociales y políticos implicados y al hecho de que tienen lugar en un contexto de mundo «real» empírico y, por lo tanto, abordan dinámicas de un mundo «real». En consecuencia, la reflexividad y la flexibilidad son esenciales no solo a la hora de desarrollar o diseñar un proyecto, sino también para la creación de marcos y medidas para evaluar los resultados e impactos de un proyecto. Dichas evaluaciones también deben tener en cuenta impactos sociales y científicos, añadiendo así otra capa más de complejidad al asunto.

Con el telón de fondo de la amplitud social y política de los dilemas implícitos en las cuatro limitaciones mencionadas anteriormente, nos gustaría terminar este documento de posición con una cuarta recomendación: *los proyectos de transdisciplinariedad con apoyo metodológico espacial para el ODS11 deben priorizar la formación de equipos de investigación y práctica que incluyan (i) miembros con sólidas competencias de metodología espacial, (ii) personas que puedan salvar la brecha entre la experiencia técnica y el conocimiento local (iii) y expertos en procesos participativos que puedan garantizar que el compromiso de las partes interesadas y de la comunidad sea significativo y sostenido.*

Index

A
Academia-school-practitioner engagement, 88–90
Academic team, 81–83, 85–87, 90, 92
Action-oriented knowledge, 24, 27
Action-research-based psychology, xix
Adaptation strategies, 125
Adi Ganga presentation, 84
Analytical hierarchy process (AHP), 122
Analytical methods, 134
Applied research, 8, 10, 11, 13, 14
ArcGIS software, 31
Argentina, 6
Ari Ecowalk, 108–112
Asia, 101
Asian Institute of Technology (AIT), 102
Ayutthaya, 106

B
Bangkok, xxii, xxiii, 99–112, 145
Bangkok Metropolitan Administration (BMA), 101–102
Bangkok urban resilience, 101
Basic research, 8, 14
Bheri/sewage-fed ponds, 81, 83
Biodiversity, 91, 92
Brazil, xxii, 42–51, 145
Bruntland report, 132
Buenos Aires, 6
Burgers Park, 138–139

C
Canonical interdisciplinarity, 31
Cape Town, 6
Central Inland Fisheries Research Institute (CIFRI), 89
Central Thailand, 105
Chicago, 61
Climate Actions Tool, 12
Climate change, xx, 101, 102, 104, 111
Climate Risk Profile Tool, 12, 13
Co-evolving mutualism, 142
Collaborative Risk-Informed Decision Analysis (CRIDA), 104
Costa Rica, xxii, 19, 28, 145
Council for Scientific and Industrial Research (CSIR), 4, 10–15
COVID-19 pandemic, xxii, 44, 52, 59
Criteria and flood resilience index, 124
Critical social science, 8
Cross-disciplinary research, 133

D
Data acquisition techniques, 118
Data analysis, 86
Deliberative reflexivity, xxiv
Digital platform
  Burgers Park, 138–139
  components, 139–140
  development, 142
  GIS, 139, 142
  Melusi Open Spaces, 138
  Moja Gabedi, 138
  Muckleneuk Commons, 136–138
  testing/launching, 140
Digital tool, 135, 139
Disappearing Dialogues (DD), 81, 82
Disaster risk management, 119, 127
Disaster risk reduction (DRR), 117, 125
Diverse methods, 147
Dried fish presentation, 84
Dry Weather Flow (DWF), 80
Durban, 4, 6, 145
2022 Durban Climate Change Strategy, 12

E
East Kolkata Wetland (EKW), 77–96
Ecosystems Services Concept, 7
Enkanini, 5
eThekwini GreenBook MetroView, 11–14
Ethno-graphers, 81, 86, 88, 92
Ethno-graphic, 81, 82, 93

Ethnographic methods/techniques, 79, 83, 84, 89
Ethnographic studies, 105, 112
Ethnography, xxiii, 61–63, 77–96, 100, 101, 105–107, 111, 112, 148
Ethno-graphy approach, 88, 92–94
Evidence-based knowledge, 146

F
Flood resilience, 117–119, 121, 122
Flood resilience assessment method, 126
Flood risk management
    community-based, 117–119
    community flood resilience, 121–125
    financial contributions, 125
    flood resilience study, 125–127
    risk perception, 119
    river basin flood susceptibility, 119–123
    spatial, 117–119
    study areas, 119
    systemic, 117–119
Flood Risk Perception Criterion, 123
Flood susceptibility, 118–122, 125–127
Flood susceptibility map, 119, 121–123, 126

G
Gauteng Province, 134
Geographic information systems (GIS), 11, 99, 101–103, 111, 112, 118, 120, 132, 139, 142
Geospatial Information Agency (BIG), 120
Geospatial methods, 118, 119
Geospatial modelling, 126
Ghana, 6
GIS Web Apps, 139–140
Gothenburg, 6
GreenBook, 11–15

H
Hanoi, 102
Harare, 6
Hatfield, 138
Holistic approach, 131
Home Environment Criterion, 123
Homelessness, 59–73, 138, 139
Hybrid methodologies, 99–112

I
India, xxii, 6, 78, 81, 93, 145
Indonesia, xxii, 117–128, 145
Indonesian Agency for Meteorology, Climatology (BMKG), 120
Innovative methods, 90, 96
Institute for Sustainable Urban Development (ISU), 7
Instrumental knowledge, 12
Interdisciplinarity, xix, 20–22, 28, 30, 31, 33
Internet, 86, 91
Intervention and evaluation research, 8, 11

J
Jakarta, xxiii, 119

K
Kenya, 6
Kharagpur, 81, 83, 85, 89
Khlongluang, 103
Kisumu, 6
Knowledge coherence, 22, 134
Knowledge unity, 22, 52, 134
Kolkata, xxii, xxiii, 77–80, 82, 83, 89, 93, 94, 145
Kumasi, 6

L
Lefebvre's pioneering approach, 146
Local Interaction Platforms, 6
Low-impact urban design and development (LIUDD), xxi
Lund, 6

M
Makhanda, 6
Malmö, 6
Melusi Open Spaces, 138
Master planning, xxiii, 28–30
Methodological coherence, 7, 10
Methodological frameworks, 8, 9
    applied research, 8, 10, 11, 13, 14
    basic research, 8, 14
    critical social science, 8
    dimensions, social science research, 8
    pragmatism, 8
    spatial methods, 8
MetroView, 11–14
Mistra Urban Futures (MUF), 6
Mode 1 knowledge production, 3
Mode 2 knowledge production, 3
Mode 2 research, xix, 3, 30
Moja Gabedi, 138
Mooca district
    street sidewalk, 69
Muckleneuk, 136
Muckleneuk Commons, 136–138
Multi-criteria evaluation approach, 122
Multidisciplinarity, 20–21, 24, 32
Multidisciplinary research, 133
Multi, inter-and transdisciplinarity, 19–22
Multi-layer perceptron Markov model, 102

N
Nalban *bheri*, 87, 88
National Housing Authority (NHA), 105
National Research Foundation (NRF), 135
Nature-based resource, 92
Nicolescuian transdisciplinarity, xix
Nonthaburi, 103

P

Participatory action research (PAR), xix, 8, 10, 38
Participatory approaches, 118, 125, 127, 128
Participatory Flood Risk Management (PERM), 125
Participatory methodology, 121
Participatory planning, 38–41
Participatory research (PR), xix, 9, 14, 38, 39, 46, 52, 53, 125
Participatory-research education, xix
Patachitra play, 93, 94
Planes reguladores, 28–30, 32
Planning support science (PSSci), 4, 11, 14
Planning support systems (PSSs), 4, 9–14
Plastic waste management, 90
Porto Alegre, xxii, xxiii, 38, 145
    actors and their roles, 47
    cultural gathering, 48
    historic waterfront development, 42–51
    steps, in transdisciplinary case study, 49–50
    transdisciplinary research process, 48
Practical Empirical Implementation Project (PEIP), 77–79, 89–92, 95
Practitioners, 90–94, 118, 145
Pragmatism, 8
Pragmatist reflexivity, xxiv
Pretoria, xxii, xxiii, 134, 138
Project
    Academia-School-Practitioner Engagement, 88–90
    design and team composition, 81–82
    ethno-graphic training/exchange/exposure, 82–86
    selections and combinations of methods, 86–88
    three-fold execution plan, 82
Prototypical designs, 6, 10, 14
Public engagement, 6
Public spaces, 3, 45, 46, 48–50, 59–61, 65–72, 131–142, 146

Q

Qualitative method, 103, 105, 122, 126
Qualitative methodology technique, 84

R

Rajarhat, 80
Ramsar Convention, 78
Reflexive knowledge, 6, 8, 9, 12, 14
Regenerative Public Space (RPS)
    blurring of boundaries, collaboration/participation/cooperation., 140–143
    digital platform, 136–140
    framework, 135–136
    phases, 141–142
    study areas, 137
    Tshwane, 134–135
Regenerative sustainability, 131–134, 140, 142
Remote sensing, 102, 118, 127
Research design, 10, 13, 39
Research methods, 9, 10, 23, 33, 34, 38, 62, 85, 92, 101, 105, 132, 134, 135, 140, 142
Resilience, xxi, xxv, 12, 60, 78, 100–107, 111, 112, 117–119, 121–128, 131–133, 146
Risk perception, 119

S

Sagar Islands, 83
San José, xxii, xxiii, 145
São Paulo, xxii, xxiii, 145
    COVID-19, 59–61, 66, 72
    homelessness, 62, 64, 65, 68, 71
    MEPSRSP, 64
Science communication, 6
Scotland, 121
Sheffield and Manchester (UK), 6
Shimla, 6
SMUS Toolkit, 59, 61–65
Social inclusion, 68, 69
Social interactions, 65, 66, 147
Social learning, 19, 147, 148
    definition, 19–20
    deliberative reflexivity, 25, 26
    interactive and purposive, 20
    multi-, inter-and transdisciplinarity, 20–21
    as planning practice and urban development catalyser, 24–25
    pragmatist reflexivity, 25, 26
    substantive and instrumental, 20
    transdisciplinary research, 24
Social Polis, 4–5
Social science research, 8
Socio-ecological context, 91
Socio-ecological systems (SES), 132, 133, 142
South Africa, xxii, 4–6, 12, 14, 131–143, 145
Spatial analysis, 103–105, 111, 112, 120, 122
Spatial approach, 101, 118, 127
Spatial data, 99
Spatially sensitive methods, 146, 147
Spatial-methodological toolkit, 146
Spatial methods, xx, 4, 8, 79, 88, 95, 99, 119, 127, 132, 145
    Bangkok urban resilience, 101
    combination, SDG11, 147
    dynamic participatory and transdisciplinary process, 46
    ethnography, 105–107, 111, 112
    GIS, 101–103, 111, 112
    limitations, 148–149
    methodologies, 100
    possibilities, 147–148
    sensorial and affective methods, 108–111
    SMUS, xxi
    spatial analysis, 103–105, 111, 112
    spatial-methodological transdisciplinarity, xxii
    transdisciplinary projects, 147
    transformative role, xxiii–xxiv, 146–147
    urban studies, 100
    uses, 107–108, 111
    utilisation, 100
    visual tools, 100

Spatial Methods for Urban Sustainability (SMUS), 77–79, 81, 89, 90, 95
Spatial research, xxi, xxiii, xxiv, 23, 33, 34, 41, 42, 62, 100, 107, 147
Special Social Approach Service (SEAS), 64
Stakeholder engagement, 6, 41
Stakeholders, 107, 108, 118, 119, 127, 132, 141
Stockholm, 6
Storm Water Flow (SWF), 80
Students, xxiii, 41, 44, 45, 47, 49, 50, 62–64, 79, 81–95, 138, 148
Sustainability research, xx, 8, 33, 107, 134
Sustainability thinking and implications, 132–134
Sustainable development, 60, 132
Sustainable Development Goal 11 (SDG11), xix–xxv, 3, 24, 37–38, 43–46, 52, 77–79, 101, 107, 131, 145–149
   goals, 52
   and homelessness, 59–73
Sustainable Development Goals (SDGs), 3, 6, 24, 44, 52, 60, 78, 131, 135
Sweden, 6

T
Tangerang, xxii, 117–128, 145
Thailand, xxii, 106, 111, 145
Topographic Wetness Index (TWI), 120
Toronto, 26
Traditional methods, 147
Transdisciplinarity, 95, 96, 107
   definition, 3–4
   multi-, inter-and transdisciplinarity, 20–21
   participation in research and practice, 38–40
   in practice, 22–23
   research and practice, 19–34
   for urban sustainability
      African, 5–6
      cross-country, 6–7
      institutional, 7
      Social Polis, 4–5
Transdisciplinary approaches, xxii, 5, 72, 99, 117–119, 121, 125–127, 133, 134, 140, 142, 143
Transdisciplinary participatory research, 37–53
Transdisciplinary projects, 146–149
Transdisciplinary research, xx, xxi, xxiv, xxv, 4–7, 10, 14, 15, 19–34, 38–40, 42–52, 62, 101, 108, 131–134, 147, 148
Transdisciplinary urban planning, 4, 7–11, 14, 15, 38, 147
Transformations, xxiv, 5, 8, 9, 26, 31, 44, 103, 131, 132, 135, 136, 147
Transect walk method, 83, 86
Tshwane, xxii, xxiii, 132, 134–136, 138–140, 142, 145
TU Berlin, 38

U
Urban design methods, 81
Urban development policies, 38, 44, 45
Urban planning, 3, 41
Urban studies, 99
Urban sustainability, xix–xxv, 4–8, 12, 14, 24, 59–64, 66, 67, 69, 73, 77, 89, 99–112, 127, 131, 132, 134, 142, 147
Urban sustainability (SDG #11), 59–73

V
Vietnam, 126
Visualisation techniques, xxi, 77, 79, 89
Visual methods, 78

W
Wastewater mechanism, 86, 87
West Bengal, 81
Wetlands, 78, 80, 93
WhatsApp, 62, 65, 87
'Wicked problems,' xx, 22, 26, 34

Y
'Youth for Climate' programme, 92

Z
Zimbabwe, 6
Zurich School, xix

The manufacturer's authorised representative in the EU is Springer Nature Customer Service Centre GmbH, Europaplatz 3, 69115 Heidelberg, Germany. If you have any concerns regarding our products, please contact ProductSafety@springernature.com

Printed and bound by CPI Group (UK) Ltd, Croydon, CR0 4YY

26/03/2026

02078941-0019